Study Guide for

Physical Geology

Exploring the Earth

Third Edition

James S. Monroe
Central Michigan University

Reed Wicander
Central Michigan University

Prepared by
Christopher Haley
Virginia Wesleyan College

West/Wadsworth
I⊤P® An International Thomson Publishing Company

Belmont, CA • Albany, NY • Bonn • Boston • Cincinnati • Detroit • Johannesburg • London
Madrid • Melbourne • Mexico City • New York • Paris • Singapore • Tokyo • Toronto • Washington

Printed in the United States of America
1 2 3 4 5 6 7 8 9 10

For more information, contact Wadsworth Publishing Company, 10 Davis Drive, Belmont, CA 94002, or
electronically at http://www.thomson.com/wadsworth.html

International Thomson Publishing Europe
Berkshire House
168-173 High Holborn
London, WC1V 7AA, United Kingdom

International Thomson Editores
Seneca, 53
Colonia Polanco
11560 México D.F. México

Nelson ITP, Australia
102 Dodds Street
South Melbourne
Victoria 3205 Australia

International Thomson Publishing Asia
60 Albert Street #15-01
Albert Complex
Singapore 189969

Nelson Canada
1120 Birchmount Road
Scarborough, Ontario
Canada M1K 5G4

International Thomson Publishing Japan
Hirakawa-cho Kyowa Building, 3F
2-2-1 Hirakawa-cho, Chiyoda-ku
Chiyoda-ku, Tokyo 102, Japan

International Thomson Publishing Southern Africa
Building 18, Constantia Square
138 Sixteenth Road, P.O. Box 2459
Halfway House, 1685 South Africa

ISBN 0-534-53776-6

PREFACE

The purpose of this study guide is to assist the college student in the introductory level geology class with the sometimes daunting task of assimilating the material presented in the textbook. It is designed both to act as a guide while reading each chapter and to test you on the material after the chapter has been read and studied. To maximize its use read this preface before starting the first chapter.

STRATEGIES FOR SUCCESS. A SPECIAL INTRODUCTION FOR FIRST YEAR COLLEGE STUDENTS.

The first year of college is always an exciting experience, but often a frustrating one as well. For most it is the first time you will make many decisions that were made for you by others in the past. You now have to decide when to do homework, when to watch TV, when to go to parties, whether or not to go to class. While the freedom is refreshing, the distractions are many and many college freshmen find themselves not doing as well academically as they thought they would. It is important to be aware that the level of achievement expected for a given grade in college is generally much higher than any previously encountered. Listed below are some suggestions on avoiding the pitfalls that many students fall into their first year.

1. GO TO CLASS, EVERY CLASS. In secondary school this was not a decision normally left up to you. Although, there are no truant officers to make you attend class in college, I would suggest that any perception that you can get away with skipping class is an illusion. Skipping classes is the single biggest mistake that college students make their first year, and continue to make in subsequent years if they fail to recognize it as a problem. I would submit that skipping classes is a drug just as surely as alcohol or illegal narcotics. It is addicting and destructive yet can be so easily rationalized in the mind of the individual that he or she is positive that the reasons for a bad grade in the class are everything *but* the failure to attend.

Some college instructors will deduct from your grade if you skip too many classes and they say so at the beginning of the semester. By and large, college administrations will back up an instructor's decision to deduct from a grade for this reason, so it is useless to appeal. On the other hand, many instructors do not require class attendance. These instructors feel that the grade achieved will probably reflect this lack of attendance and that a stated requirement is not necessary. While this may seem devious, it is also correct. In the "real world" there are dire consequences for skipping work. So, too, in college.

2. LEARN TO TAKE GOOD NOTES. There are several reasons for this. Most importantly, no instructor has exactly the same vision of what is important as the authors of your textbook. Almost all instructors at some point present material that is not emphasized in the book, and for this your notes will be crucial. Second, your instructor may have a different way of explaining concepts also presented in the book. The more different ways that you are presented material, the better the chances of your grasping the important concepts. Third, taking notes keeps you "in the ball game". In large classes, especially, where student participation can easily be avoided it is easy to start to notice things like chalk dust on the instructor's clothing, the student constantly sniffing to your left, the clown behind you making wisecracks. Concentration is difficult, but taking notes can help.

Good and useful notes are aided greatly by sitting toward the front of the classroom. You can see and hear better and there are fewer distractions. Another good habit is to copy your notes over at the end of class. This makes them neater and gives you a chance to translate the "short hand" symbols in your notes that "you were sure you would remember what they meant".

3. SHOW UP IN CLASS PREPARED. Generally this means reading the assignment before it is discussed in lecture. The text in your book is very concise. You will rarely, if ever, have to read more than one chapter before a lecture. This can be achieved in an hour or two for most of the chapters in

your book. Having been introduced to vocabulary before a lecture makes the lecture clearer and more interesting. Study the diagrams and figures in your chapter before class as well. Geology is a very visual subject and the diagrams in your textbook are, indeed, "worth a thousand words".

4. ASK QUESTIONS AND UTILIZE YOUR INSTRUCTOR'S OFFICE HOURS. There is no point in beating your head against the wall if there is some information that you just can't seem to assimilate. Many students have one major mistaken perception about asking questions: that the instructor and other students in the class will think the question (and therefore the student) is dumb. Forget about it. If you are lost there is a good chance that you are not alone. Think about this, when was the last time that *you* thought that a fellow student was an idiot for speaking up and asking a question?

Nevertheless, if asking questions in the public forum of 100 or more students is not your style then use the instructor's office hours. These one-on-one encounters may pleasantly surprise you. For instance, you may find out that your instructor is a human being with concern and compassion for the student (it's been known to happen). You may find out that you are having trouble with material because you don't understand something else that came before, thereby avoiding a disaster. You actually may get to know the instructor which, somehow, often seems to help. Not enough students take advantage of office hours. Second only to your own mind, the best resources you have at college are the minds of your instructors. Don't fail to take advantage of that.

TIPS ON STUDYING FOR EXAMS

1. DON'T WAIT UNTIL THE LAST MINUTE. You need to leave ample time to get help from the instructor (see #4 above) if you suddenly discover that there is something that you don't understand. Waiting until the last minute inevitably leads to the "all nighter". Staying up all night before a test usually has the same effect on your performance that it would have on a Marathon runner's performance. Rest the night before the event is essential.

2. STUDY IN A GROUP, BUT WITH THESE GUIDELINES: Do not depend on the group to get you through the exam. They won't be there for you when you take the exam. Make sure your group stays focused on the task at hand. Discussing sports, fashion, relationships, the food at the dining hall, or which fraternity or sorority you are thinking about rushing can wait. If your study mates insist on talking about anything except geology, walk away. Leave some time to study alone after your group meets. Groups tend to decide what to go over by committee and this may not be what you need to work on most.

3. USE OLD EXAMS IF THEY ARE AVAILABLE. These might be useful if your instructor has put them on file (not that many do). However, you should only use them as a guide for the type of material that was emphasized and the types of questions asked. DO NOT memorize the questions and hope that they will be the same on your exam. If an instructor gives the same test year in and year out, you can bet that the test is not on file.

4. USE THIS STUDY GUIDE. Treat the questions in this study guide and those that follow each chapter in your textbook as sample tests. Some hints on how to maximize the use of this guide are given below.

USE OF THIS STUDY GUIDE

This study guide is meant to be a supplement to your textbook and notes. You should be under no delusion that use of this guide is in any way replacement for attending class and reading your text. That said, however, you should find it a very helpful guide to the information in your textbook. Each chapter is divided into six to nine parts. These include: 1) Chapter Objectives, 2) Key Terms, 3) Chapter Concept Questions, 4) Complete the Table (not in all chapters), 5) Completion Questions, 6) Multiple Choice, 7)

True or False, 8) Drawings and Figures (in most, but not all chapters) and 9) Useful Analogies (not in all chapters).

CHAPTER OBJECTIVES

This is a list of goals that you should set for yourself when attacking a chapter. They are listed in the same order as the information that should help you achieve these goals is presented in the textbook. To maximize the use of this section you should read it over once before you start the chapter, then keep it in front of you while you read the chapter. When you finish reading a section of the chapter, read the objectives that fit the information covered in that section. Satisfy yourself that you can fulfill that objective or at least that you understand what that objective entails. If you do not feel confident that you have achieved the objectives for some sections, be sure to reread those sections.

KEY TERMS

This is a list of geologic terms that are introduced in each chapter. Half the battle in conquering the subject of geology is mastering the terms. Not only should you know the definition of each term but you should also be able to use the terms. You should also know why each is significant. They are listed approximately in the order that they are discussed in the text. It is helpful to keep this list in front of you as you read the chapter, highlighting the terms as they are encountered in the text. When studying for exams you can use this list as a list of terms to define.

USEFUL ANALOGIES

Some concepts are just plain difficult to understand. Although the authors of your textbook and your instructor have worked hard at explaining these concepts clearly, sometimes they just don't stick. This section attempts to explain some of the more difficult concepts in something other than geologic terms. This use of analogy may prove useful for some students. Check the end of each chapter in this manual to see if such a section is included. These are just alternative explanations rather than questions and answers.

CHAPTER CONCEPT QUESTIONS

These are discussion questions that require some thought and organization to answer. Since the wording of each answer is a highly individual matter, there was no attempt to give a specific answer in the back of the book. Instead textbook page numbers are given in the Answer appendix where information to help you answer the question may be found.

COMPLETE THE TABLE

These are included in chapters where some sort of classification is encountered. They are generally self explanatory. Filling in the required blanks may prove helpful in learning these classification systems. Like the Chapter Concept Questions, page numbers are given in lieu of exact answers.

CHAPTER COMPLETION, MULTIPLE CHOICE AND TRUE OR FALSE QUESTIONS.

These are sample questions in three formats that are commonly encountered on exams. The answers are given in the back of this manual.

DRAWINGS AND FIGURES

Geology is a very visual subject. This fact is made evident by the lavish use of diagrams and photographs in your textbook. Undoubtedly your lecturer makes use of slides and/or the videodisk that West Publishing Company has developed to accompany your textbook. Some instructors incorporate illustrations into their tests or ask questions in which you are asked to sketch. To give you practice answering these types of questions, most chapters include a section of this type.

OTHER RESOURCES

In addition to your textbook, professor and notes there may be other resources available to you at your school that you should be aware of. For example, does your school have a mineral museum? If so, twenty minutes walking around it may prove one of the most productive ways to add to your knowledge or clarify information presented in class. As previously stated, geology is a very visual field. If a picture is worth a thousand words, then actual specimens are worth ten thousand.

If you are required to write out-of-class essays or papers, or if you are just interested in learning more about this field, some of the following suggestions might prove useful.

PERIODICALS

There is a plethora of professional journals in geology for virtually every discipline within the science. Most of these will prove too esoteric or complex for the average introductory student, but if you would like to have a look, your school library probably subscribes to several. Rather than list them here (there are way too many) I would suggest going and looking in your library. The library subject catalog or the reference librarian can help you find them.

There are other periodicals that are not so directed to the professional geologist that often carry geology-related articles. *Earth* is a periodical devoted exclusively to geology and the environment. It always has nice pictures and diagrams and is specifically oriented to the high school or introductory level college student and interested layman (that's you!). It is growing in popularity and can be found in many public libraries and newsstands. *National Geographic* often has articles pertinent to environmental issues, paleontology, and even geology. *Discover* is a good college level general science publication which often has articles related to the earth sciences. *Scientific American* also is oriented toward science students. Its articles are often a bit more technical than the others listed but should not be beyond the comprehension of a college freshman. The National Parks Conservation Association publishes *National Parks,* a magazine devoted to informing the American public about one of our greatest legacies. Their articles are very informative and often contain a great deal of information on geologic features of our parks. This list is, of course, incomplete, so ask your librarian about other periodicals.

VIDEOS AND TELEVISION

Cable television and public television often carry programs devoted to the earth. The Discovery Channel has built a reputation featuring quality science oriented shows and is carried by most cable operators. Similarly, public television often has specials or even special series dedicated to the earth sciences. Nova, a weekly science program on most public television stations, offers very high quality entertainment of interest to the earth science student. New programs and cable channels are coming along all the time so the only advise I can give is to check your local listings.

You can also check your library for videos on earth science topics. Many libraries buy videos of some of the Nova programs. Three particularly good ones are <u>Volcano!</u>, <u>In the Path of a Killer Volcano</u>, and <u>A Trip Down the Grand Canyon</u>. Scott Resources of Fort Collins, Colorado also produces a series of very good earth science videos covering a wide array of topics discussed in your textbook.

These are oriented toward high school students but introductory college students may find them good supplements to reinforce what they have learned. The excellent and popular series <u>Earth Revealed</u> is also available as a video series and is probably available in your library. There are many others, too numerous to mention so, once again, check your libraries, both college and public to see what they have.

ROADSIDE GEOLOGY GUIDES

The Mountain Press publishes guidebooks for those interested in learning about the geology of the places they drive through. They are written at a level that really requires no previous knowledge but are most useful for people that have had an introductory course (that is you, almost!). One is published per state (although some states such as Vermont and New Hampshire are grouped together). At this writing, not all states are covered by the series. Many western states are completed as well as a few eastern states, but no mid-western states. It is worth finding out if your state is covered. Think how impressed your instructor will be if you write a paper on the local geology and actually visit various places in your area! To find out if your state is covered by a roadside geology guide you can contact Mountain Press Publishing Company at:

     P.O. Box 2399
     Missoula, MT 59809
     Phone: (406) 728-1900

THE INTERNET

The most rapidly growing source of information available to students in the last five years is the Internet. Virtually every college has ready access to it in the library and an increasing number of schools are wiring college dormitories so that all students can access the "web" minimum effort. Finding information has become extremely easy in the past few years with the advent of web browsers such as Yahoo, Infoseek, Webcrawler and numerous others. Unlike just a few years ago nearly every entering freshman has some experience using the Internet. For those that do not have experience, any college with access has the resources to teach the student how to use these easy search engines.

     The greatest problem with using the Internet these days in evaluating what information is good and what is unreliable. For example, any attempt to find information on fossils, earth history, or evolution is bound to result in a large number of pages created by "creation scientists". While this is not the place to debate the validity of this view of Earth, chances are that a term paper filled with such references will not impress the professor grading your term paper. The safest bet is to refer to the Wadsworth publishing page found at *http://www.wadsworth.com/geo* . Sites listed here have been screened by people qualified to evaluate the veracity of web sites. If you have any doubt when using the Internet ask your professor or lab instructor if a particular site contains reliable information.

CHAPTER 1

UNDERSTANDING THE EARTH—AN INTRODUCTION TO PHYSICAL GEOLOGY

CHAPTER OBJECTIVES

By the end of this chapter you should be able to:

1. Describe several professional endeavors in which geologists are involved.
2. Discuss how some knowledge of geology and the earth can help any citizen as a consumer and a voter.
3. Summarize the currently accepted theory for the origin of the solar system and Earth.
4. Name, compare, and contrast the different layers of the Earth.
5. Compare and contrast the two types of crust.
6. Summarize the scientific method and differentiate between a hypothesis and a theory.
7. Distinguish between the different types of plate boundaries and describe the basic characteristics of each.
8. Discuss the relationship among the three kinds of rocks and the various processes operating on and within Earth which transfer materials from one kind of rock to another.
9. Explain the principle of uniformitarianism.

KEY TERMS

After reading and studying this chapter you should know the following terms:

geology	asthenosphere
physical geology	hypothesis theory
historical geology	scientific method
sustainable development	plate tectonics
solar nebula	convection cells
planetesimal	divergent plate boundaries
mantle	converging plate boundaries
core	Pangaea
crust	subduction zone
lithosphere	transform fault

rock cycle

igneous rock

sedimentary rock

metamorphic rock

geologic time scale

uniformitarianism

CHAPTER CONCEPT QUESTIONS

1. List and define 4 fields of specialization in geology.

2. Describe three ways in which geology bears on everyday life?

3. How does *sustainable development* differ from past practices? Why is it important to the future of our society?

4. According to the most widely accepted theory, how did our solar system form?

5. Why are the heaviest elements concentrated in the deep interior of Earth?

6. Describe how scientists define a question, carry out studies and draw conclusions?

7. List the three types of plate boundaries and describe their differences in terms of relative plate motion.

8. What do geologist think is the mechanism by which plates move?

9. Using plate tectonics and the rock cycle as examples explain what is meant by the chapter title, "Dynamic Earth."

10. Although plate tectonics is essentially a phenomenon of the solid Earth, it impacts other parts of our planet. Explain how plate tectonics effects the atmosphere, hydrosphere, and biosphere.

11. What is the general origin of the igneous, sedimentary, and metamorphic rocks?

12. James Hutton's principle of Uniformitarianism is often stated as, "The present is the key to the past." Explain this statement.

COMPLETE THE FOLLOWING TABLES:

Layer	% Earth	Avg. Density	Composition
Core			
Mantle			
Crust			

Crust Type	Thickness	Avg. Density	Composition
Continental			
Oceanic			

COMPLETION QUESTIONS

1. Mineralogy is the study of _____ while petrology is the study of _____.

2. Stratigraphy is the study of _____.

3. Structural geology is the study of the _____ of the crust.

4. Paleontology is the study of _____.

5. The scientific approach of investigation involves developing a tentative explanation or _____. If this explanation is shown to be repeatedly correct, a _____ is proposed to explain the phenomena.

6. The Earth formed _____ billion years ago.

7. Our solar system formed from a rotating disk-shaped cloud of gas and dust called a _____.

8. Most of the matter in this cloud condensed in the center to form the _____, but local eddies in the disk accumulated cold matter to form the early _____.

9. The mechanism for moving the plates across the Earth appears to be _____ within the _____.

10. At divergent margins the plates are moving _____ each other.

MULTIPLE CHOICE

1. Which of the following is a pursuit in which professional geologist are employed?
    a. the search for fossil fuels
    b. the search for groundwater
    c. locating safe building sites in earthquake prone areas
    d. all of the above

2. Petrology is the study of
    a. petroleum.
    b. rocks.
    c. fossils.
    d. none of the above.

3. Paleontology is the study of
    a. ancient landforms.
    b. ancient rocks.
    c. ancient minerals.
    d. none of these.

4. During formation of our solar system ___ % of the mass in the collapsing nebulae concentrated in the center to become the sun.
    a. <10
    b. 30
    c. 60
    d. 90

5. The differentiation of the earth resulted in the formation of
    a. the layers of the earth.
    b. the atmosphere.
    c. the oceans.
    d. all of these.

6. The continental crust is rich in
    a. silicon and magnesium.
    b. aluminum and magnesium.
    c. silicon and aluminum.
    d. iron and magnesium.

7. The mantle is rich in
    a. silicon and magnesium.
    b. aluminum and magnesium.
    c. silicon and aluminum.
    d. iron and magnesium.

8. The core of the Earth is rich in
    a. iron and magnesium.
    b. iron and nickel.
    c. silicon and aluminum.
    d. silicon and oxygen.

9. The asthenosphere is composed of
    a. crust only.
    b. crust and the upper part of the mantle.
    c. crust and the lithosphere.
    d. part of the mantle only.

10. The lithosphere is composed of the
    a. crust only.
    b. crust and the entire mantle.
    c. crust and the asthenosphere.
    d. crust and the mantle above the asthenosphere.

11. Sea floor forms at a(n)
    a. deep sea trench.
    b. island chain.
    c. mountain range.
    d. mid-ocean ridge.

12. Features found at converging boundaries include
    a. transform faults.
    b. ridges.
    c. trenches.
    d. all of these.

13. The well known San Andreas Fault is a ___ boundary.
    a. divergent
    b. convergent
    c. transform fault
    d. none of these

14. Along a transform margin oceanic crust is
    a. created
    b. consumed
    c. both
    d. neither

15. The hypothesis of sea floor spreading was an improvement over continental drift because it better explained
    a. why the Africa and South America coastlines fit together.
    b. why tropical fossils occur in North America.
    c. the mechanism by which the continents can move.
    d. why the earth is a differentiated planet.

16. Igneous rocks form from
    a. pre-existing rocks undergoing changes due to heat, pressure, and chemically active fluids.
    b. a cooling magma or lava.
    c. sediments compacted and cemented.
    d. all of these

17. Absolute ages have been assigned to the geologic time scale since the discovery of:
    a. plate tectonic theory.
    b. the first fossils.
    c. radioactivity.
    d. uniformitarianism.

TRUE OR FALSE

___1.   Most of the energy resources now being consumed are non-renewable.

___2.   Most of the material from the solar nebula, that formed our solar system, is found in the planets.

___3.   The Earth is composed of layers of different thicknesses and composition.

___4.   The outer core is the thickest layer of the earth.

___5.   The asthenosphere rests above the lithosphere.

___6.   The oceanic crust is denser than the continental crust.

___7.   Continental crust is thicker than oceanic crust.

___8.   The core is denser than the mantle.

___9.   The asthenosphere includes the crust and the upper most part of the mantle.

___10.   A hypothesis can only be developed, but it can not be tested.

___11.   Subduction zones or trenches form as an ocean bearing plate sinks beneath a continental bearing plate.

___12.   The hydrosphere and atmosphere of Earth are independent of the solid Earth and are unaffected by Plate tectonics.

___13.   Sedimentary rocks form from the melting of pre-existing rocks.

___14.   Metamorphic rocks cannot undergo weathering.

___15.   Present day geologic processes have always operated at a uniform rate.

DRAWINGS AND FIGURES

1.   Reproduce the rock cycle as illustrate in Figure 1-17. Be sure to show all possible paths and label rock types and processes. Your version doesn't need to be as fancy as that in the book, but it should be accurate.

2. In the figure below label the layers of the Earth. In the inset in the upper right label the lithosphere, asthenosphere, upper mantle and the two types of crust. Check your answer against Figure 1-13 in your textbook.

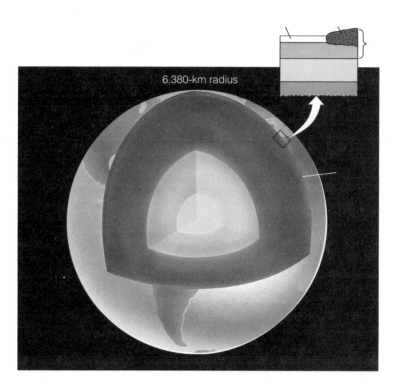

3. In the diagram below label each plate boundary as to type. For convergent boundaries be sure to indicate type of crust (e.g. continent-continent). Also indicate where subduction zones, trenches and mid-ocean ridges occur. The correct answer is Figure 1-16 in your textbook.

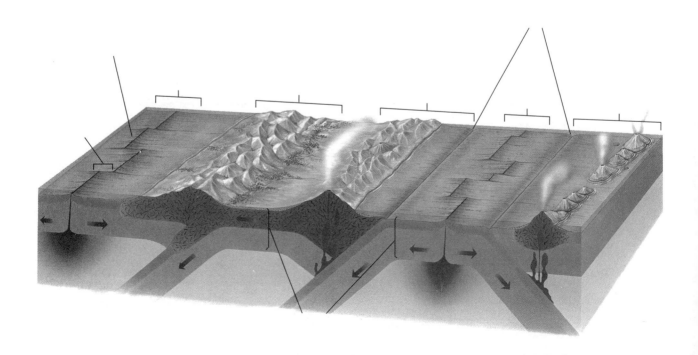

CHAPTER 2

MINERALS

CHAPTER OBJECTIVES

By the end of this chapter you should be able to:

1. Describe the parts of an atom and explain why some tend to ionize.
2. Describe the various types of bonds found in minerals.
3. Fully define "mineral" as defined by geologists and, for a variety of substances, say whether or not they are minerals and why.
4. List the common elements of Earth's crust in order of abundance.
5. List the common mineral groups and their characteristic compositions.
6. Explain how silicate minerals are put together.
7. Be familiar with the names of some of the most common minerals and be able to place them in the proper groups.
8. List and define several physical properties of minerals and know what controls them.
9. Differentiate between a rock and a mineral.
10. Describe several ways in which minerals have artistic and economic value.

A USEFUL ANALOGY

A simple model illustrating the effect of bond type (and, therefore, strength) on cleavage can be built by pasting several pencils together using classroom type paste. The pencils represent a chain of silica tetrahedra covalently bonded. The paste can be considered ionic bonds holding the chains together. Make several layers in this way. If you drop the model it will always break the pasted bonds (i.e. it will always break the bonds holding the pencils together) rather than breaking the pencils themselves. Although it is possible to break the pencils (the covalent bonds) it is much easier to break the pencils apart. Furthermore, when you break the pencils apart they break along plains (cleavage!) and when you break across the pencils the break is irregular (a lack of cleavage!).

KEY TERMS

After reading and studying this chapter you should know the following terms:

mineral
elements
atoms
nucleus
protons
neutrons
electrons
electron shells
atomic number
atomic mass number
noble gas
isotopes
bonding
compound
ion
ionic bond
covalent bond
metallic bond
van der Waals bond
naturally occurring
inorganic
crystalline solid

amorphous
interfacial angle
native elements
silica
silicates
silica tetrahedron
ferromagnesian
nonferromagnesian
carbonates
luster
crystal form
cleavage
fracture
hardness
Mohs hardness scale
specific gravity
polymorphs
gemstone
rock-forming minerals
resource
reserve

CHAPTER CONCEPT QUESTIONS

1. Completely define the term *mineral*.

2. How are atomic number and mass determined?

3. Is sugar a mineral? How about water or ice? For each explain why or why not.

4. Distinguish between an element and an isotope in terms of the nucleus of an atom.

5. If diamonds and graphite are both composed of carbon atoms, why are they different minerals?

6. Why do many atoms tend to become ions?

7. Describe several ways in which atoms join together in compounds?

8. What is meant by "constancy of interfacial angles?"

9. Considering the large number of elements in existence and number of combinations possible, why is Earth composed of relatively few abundant minerals?

10. Look at Figure 2-9. In minerals it is common for calcium ions to substitute for sodium ions and for magnesium ions to substitute for iron ions without substantially changing the crystal structure. However, iron never substitutes for calcium or sodium. Why?

11. List some characteristics that are used to identify minerals. For each, describe factors (composition, crystal structure, etc.) that determine that particular characteristic.

12. Why is it uncommon for minerals to display perfect crystal forms. What factors allow a crystal of a mineral to display its perfect crystal form?

13. What does a specific gravity of 2.3 mean?

14. List three ways in which minerals form.

15. Under what circumstances is a mineral a rock-forming mineral?

16. What is the difference between mineral resources and mineral reserves?

COMPLETION QUESTIONS

1. To be considered a mineral, a substance must be _____ occurring and be a _____ solid (i.e. it must have a specific internal structure).

2. _____ are aggregates of one or more minerals.

3. Anything that has mass and occupies space is considered _____.

4. All matter is composed of chemical _____, which are composed of particles called _____.

5. The central portion of an atom is called the _____, in which there are positively charged _____ and neutrally charged _____.

6. The number of protons in the nucleus of an atom determines its _____ number.

7. The atomic mass number is the sum of the number of _____ and _____ present.

8. Isotopes of an element all have the same number of _____, but a different number of _____ in the nucleus.

9. The process of joining atoms together is called _____.

10. When atoms of two or more different elements are joined together the resulting substance is a _____.

11. When an atom gains or loses an electron, a charged particle called an _____ forms.

12. The attractive force between ions of opposite electrical charges produces _____ bonding. When electrons are shared by adjacent atoms , _____ bonding occurs.

13. When the electrons of the outermost shell move freely from one atom to another, the atoms have _____ bonding.

14. Minerals that are only composed of one element are called _____.

15. Minerals are classified into groups that have the same negatively charged _____.

16. The two most abundant elements in Earth's crust are _____ and _____ which combine with other elements to form the group of minerals known as the _____.

17. The ferromagnesian silicates form _____ colored minerals.

18. Calcite is one of the most abundant non-silicate minerals. It belongs to the _____ group.

19. Minerals are most commonly identified by using their _____ properties.

20. The appearance of a mineral in reflected light determines its _____.

21. The ability of a mineral to break along smooth planes is called _____ but when a mineral breaks along random, irregular surfaces it is showing _____.

22. The resistance of a mineral to abrasion determines its _____.

23. Specific gravity is the ratio of the weight of a mineral compared to the weight of an _____ volume of _____.

24. The most abundant minerals which make up the bulk of Earth's crust are called the _____ forming minerals. Most belong to the _____ group of minerals.

MULTIPLE CHOICE

1. Minerals are
   a. naturally occurring.
   b. inorganic in their composition.
   c. crystalline solids.
   d. all of the above.

2. The smallest unit of matter that retains the characteristics of an element is an
   a. ion.
   b. isotope.
   c. atom.
   d. electron.

3. Protons have an electric charge that is
   a. negative.
   b. positive.
   c. neutral.
   d. none of the above

4. The atomic number is determined by the
   a. number of neutrons present.
   b. number of electrons present.
   c. number of protons present.
   d. number of protons plus the number of electrons present.

5. Except for the innermost shell, electron shells can hold up to ___ electrons.
   a. 2
   b. 6
   c. 8
   d. 12

6. An element that has a full outer electron shell when electrically neutral is called a(n)
   a. ion.
   b. isotope.
   c. noble gas.
   d. native element.

7. A weak attractive force between electrically neutral atoms results in
   a. covalent bonds.
   b. ionic bonds.
   c. van der Waals bonds.
   d. metallic bonds.

8. Quartz is composed of
   a. calcium carbonate.
   b. silicon and oxygen.
   c. potassium and silicon.
   d. none of these.

9. Minerals composed of atoms of a single type are called
   a. oxide minerals.
   b. silicate minerals.
   c. carbonate minerals.
   d. none of these.

10. Strong bonds in which electrons are shared between adjacent atoms are
    a. ionic.
    b. covalent.
    c. metallic.
    d. van der Waals.

11. If a mineral has a high electrical conductivity it most likely has a(n) ___ bond.
    a. ionic
    b. covalent
    c. metallic
    d. van der Waals

12. When a mineral breaks along smooth planes, it has
    a. fracture.
    b. streak.
    c. cleavage.
    d. all of these.

13. If a substance is constructed of atoms arranged in a regular three-dimensional arrangement it is
    a. crystalline.
    b. a glass.
    c. amorphous.
    d. none of the above

14. The most abundant minerals are the
    a. carbonates.
    b. sulfates.
    c. silicates.
    d. sulfides.

15. Ferromagnesian silicates are rich in
    a. iron and manganese.
    b. iron and magnesium.
    c. silicon and aluminum.
    d. none of the above

16. The charge on a silica tetrahedron unlinked to any other silica tetrahedra is
    a. +2
    b. -2
    c. -4
    d. +4

17. The most common minerals in Earth's crust belong to a group known collectively as
    a. quartz.
    b. biotite.
    c. augite.
    d. feldspar.

18. A very abundant carbonate mineral is
    a. gypsum.
    b. halite.
    c. calcite.
    d. hematite.

19. Which physical property is controlled by structure and bond strength?
    a. color
    b. luster
    c. cleavage
    d. specific gravity

20. What is the most common element in the Earth's crust?
    a. iron
    b. silicon
    c. aluminum
    d. none of these

TRUE OR FALSE

___1.   Minerals are assemblages of rocks.

___2.   Electrons have a negative charge.

___3.   The atomic mass number is determined by the sum of the number of protons and the number of neutrons present.

___4.   Covalent bonding is due to the sharing of electrons.

___5.   Minerals are always crystalline solids.

___6.   Two crystals of the same mineral will generally exhibit identical angle between similar crystal faces, even if they are of very different sizes.

___7.   Silica is composed of silicon and carbon.

___8.   The basic building block of the silicate minerals is tetragonal in shape.

___9.   Nonferromagnesian silicates are dark in color.

___10.   Quartz will scratch glass.

___11.   Calcite is the most common carbonate mineral.

___12.   Hematite is a very important ore of copper.

___13.   Color is the most diagnostic of all physical properties.

___14.   Fracture is mineral breakage along random, irregular surfaces.

___15.   All minerals have cleavage.

___16.   Abundance is an important criterion for characterizing a mineral as *rock forming*.

___17.   A resource is the total amount of a commodity whether discovered or undiscovered.

___18.   A reserve is the total amount of a commodity whether discovered or undiscovered.

___19.   Ferromagnesian minerals generally have a higher specific gravity than non ferromagnesian minerals.

___20.   Most mineral resources are renewable.

___21.   Most of the world's mineral resources come from less than 200 locations.

DRAWING QUESTION

Sketch the structure of olivine, pyroxene, amphibole, and mica. The correct answer is provided by figure 2-11.

CHAPTER 3

IGNEOUS ROCKS AND INTRUSIVE IGNEOUS ACTIVITY

CHAPTER OBJECTIVES

By the end of this chapter you should be able to:

1. Explain the difference between felsic, intermediate, and mafic composition of rocks or magma.
2. Know what viscosity is, what controls it, and be able to explain how it affects the behavior of a magma.
3. Reproduce Bowen's Reaction series and explain how it works.
4. Explain how magma is generated at mid-ocean ridges.
5. Suggest reasons why magmas associated with subduction zone volcanism are commonly intermediate to felsic in composition.
6. Describe three ways in which an initially mafic magma can become more silica rich.
7. List and define the common textures found in igneous rocks.
8. Classify the igneous rocks using compostion and texture.
9. Classify the various types of igneous intrusions by their geometry and relationships to the surrounding country rock.
10. Be familiar with the various theories for emplacement of batholiths.

A USEFUL ANALOGY

It may not be immediately obvious why the crystallization and subsequent removal of olivine from a magma via crystal settling enriches the remaining magma in silica. Here is a simple mathematical experiment to show that it happens. Although far simpler than an actual magma, the principle is the same:

Imagine a jar filled with 100 marbles of three different colors. Fifty of the marbles are blue, 25 are green, and 25 are red. This is analogous to a magma with 50% silica (blue marbles), 25% magnesium (green), and 25% iron (red). To make an olivine you combine one of each color. To remove that olivine from the magma simply remove one marble of each color from the jar. Do this ten times then calculate how many of each color you have left. You should have 40 blue, 15 green, and 15 red for a total of 70 marbles. Whereas before 50% of the marbles were blue, after the removal of "olivine" blue marbles constitute 57% of the total. The jar is thus enriched in blue marbles (i.e. the silica)!

KEY TERMS

After reading and studying this chapter, you should know the following terms:

magma
lava
pyroclastic
extrusive
intrusive
felsic
intermediate
mafic
viscosity
magma chamber
Bowen's reaction series
continuous branch
discontinuous branch
geothermal gradient
mantle plumes
partial melting
crystal settling
assimilation
country rock
inclusion
magma mixing
aphanitic
phaneritic
porphyritic
phenocrysts
groundmass
natural glass
vesicle
vesicular

pyroclastic or fragmental texture
ultramafic
peridotite
basalt
gabbro
andesite
diorite
rhyolite
granite
granodiorite
pegmatite
tuff
welded tuff
volcanic breccia
obsidian
pumice
pluton
dike
sill
concordant
discordant
laccoliths
volcanic pipe
volcanic neck
batholith
stock
granitization
forceful injection
stoping

CHAPTER CONCEPT QUESTIONS

1. How does an extrusive igneous rock differ from an intrusive igneous rock in mode of origin and in texture?

2. What is the difference between the continuous and the discontinuous reaction series?

3. Why does crust melt below spreading ridges to form magma?

4. How do geologists explain the fact that an ultramafic mantle melts to produce a mafic magma at spreading ridges?

5. Explain two ways in which a mafic magma can become richer in silica as it moves towards the surface.

6. What does a porphyritic texture in an igneous rock tell you about the cooling history of the magma?

7. On what basis are igneous rocks classified? How do geologists generally identify igneous rocks?

8. Why are rhyolite liquid lava flows so rare in comparison to rhyolite tuffs?

9. Why do pegmatites tend to grow relatively few large crystals rather than many smaller crystals like most igneous rocks? What accounts for the large number of unusual minerals in pegmatites?

10. Pumice is both vesicular and glassy. What does this tell you about the magma from which it formed?

11. Explain the difference between a dike, sill, laccolith and stock.

12. What has the study of salt domes in sediments beneath the Gulf Coast suggested about the emplacement mechanism of plutons?

13. How does batholith emplacement by stoping differ from emplacement by forceful injection?

COMPLETE THE FOLLOWING TABLE:

	Felsic	Intermediate	Mafic	Ultramafic
% Silica				
Aphanitic rock				
Phaneritic rock				

COMPLETION QUESTIONS

1. Based on their origin, there are two types of igneous rocks, the volcanic or _____ igneous rocks, and the plutonic or _____ igneous rocks.

2. In the discontinuous branch of Bowen's reaction series, as the magma cools, _____-rich plagioclase feldspar reacts with the melt to form _____-rich plagioclase feldspar.

3. The discontinuous branch of Bowen's reaction series begins with the crystallization of _____ and ends with the crystallization of _____.

4. Temperature increases at a rate of about _____ degrees/km going downward into the crust.

5. Crystal settling, assimilation and magma mixing are processes that, to some degree, change the composition of a magma from _____ to _____.

6. The volcanic igneous rocks have crystals which are too fine to be seen with the naked eye and are said to have a_____ texture.

7. Plutonic igneous rocks have crystals coarse enough to be seen with the naked eye and are said to have a _____ texture.

8. If an igneous rock has experienced two phases of cooling, one slowly at depth and a second more quickly on the surface, then it develops a _____ texture.

9. An extrusive rock with holes, due to escaping gases, has a _____ texture.

10. Explosive volcanic eruptions produce fragments of various sizes that accumulate to form rocks with a _____ texture.

11. Igneous rocks are classified according to their _____ and mineralogical _____.

12. Felsic rocks have a silica content exceeding ____%.

13. A felsic, aphanitic igneous rock is called a _____.

14. An intermediate, phaneritic igneous is called a _____.

15. _____ is the aphanitic compositional equivalent of gabbro.

16. The most abundant mineral in ultramafic rocks is _____.

17. An igneous rock in which most crystal exceed one cm. in size is called a _____.

18. A rock composed of fine-grained ash particles is classified as a _____.

19. Most pegmatites are similar to _____ in composition.

20. _____ is generally a dark colored volcanic glass with a conchoidal fracture, whereas _____ is a vesicular volcanic glass.

21. When a magma solidifies in the subsurface, it forms a rock body called a _____.

22. A tabular, discordant pluton is called a _____ and a tabular concordant one is a _____.

23. Non-tabular, concordant plutons are called _____.

24. Devil's tower, Wyoming and Ship Rock, New Mexico are famous examples of volcanic _____.

25. The difference between a batholith and a stock is that batholiths have surface areas greater than ____ square km.

26. Most batholiths are _____ in composition, although a few are _____ in composition.

27. Most geologists believe that batholiths are emplaced by _____ _____.

28. _____ is a mechanism of batholith emplacement where blocks of country rock are engulfed by, rather than shouldered aside by rising magma.

MULTIPLE CHOICE

1. Intrusive igneous rock forms
   a. on the surface.
   b. by violent volcanic eruptions.
   c. from glassy lavas.
   d. none of these.

2. Compared to felsic magmas, mafic magmas are relatively enriched in
    a. calcium.
    b. iron.
    c. magnesium.
    d. all of the above

3. The discontinuous branch of Bowen's reaction series comprises the
    a. ferromagnesian silicates.
    b. nonferromagnesian silicates.
    c. highest silica minerals.
    d. all of these

4. The discontinuous branch of Bowen's reaction series comprises
    a. silicates made of isolated silica tetrahedra.
    b. single chain silicates.
    c. double chain silicates.
    d. all of the above

5. The continuous branch of Bowen's reaction series consists of plagioclase feldspars in which calcium ions are replaced by ____ ions during cooling
    a. potassium.
    b. magnesium.
    c. iron.
    d. none of the above

6. The last mineral to form as a magma solidifies is
    a. olivine.
    b. quartz.
    c. feldspar.
    d. muscovite.

7. Magmas formed at spreading ridges are invariably
    a. felsic.
    b. intermediate.
    c. mafic.
    d. ultramafic.

8. Partial melting of mantle or mafic crust explains
    a. the differences in the age of some igneous rocks.
    b. the differences in the size of some volcanoes.
    c. the differences in the chemical composition of magmas.
    d. none of the above.

9. The presence of inclusions of country rock in a felsic intrusive igneous rock may be evidence of
    a. crystal settling.
    b. assimilation.
    c. magma mixing.
    d. all of the above.

10. Coarse-grained or phaneritic rocks indicate
    a. slow cooling on the surface.
    b. rapid cooling at depth.
    c. slow cooling at depth.
    d. none of these

11. A glassy texture indicates
    a. very rapid cooling.
    b. slow cooling.
    c. slow cooling and rapid cooling.
    d. none of these

12. Pyroclastic igneous rocks from by
    a. a lava cooling on the surface.
    b. violent, explosive volcanic eruption.
    c. a magma cooling slowly in the subsurface.
    d. two phases of cooling, one fast and one slow.

13. The presence of phenocrysts in an igneous rock indicate
    a. cooling so rapid that crystals do not have time to form.
    b. violent, explosive volcanic eruption.
    c. the escape of gas from a magma.
    d. two phases of cooling, one fast and one slow.

14. Vesicles in an igneous rock form from
    a. escaping gases.
    b. phenocrysts.
    c. falling ash.
    d. none of these

15. An aphanitic, dark-colored igneous rock is called a
    a. basalt.
    b. gabbro.
    c. rhyolite.
    d. none of these

16. Rhyolite is a(n)
    a. phaneritic intermediate igneous rock.
    b. aphanitic, mafic igneous rock.
    c. phaneritic, felsic igneous rock.
    d. aphanitic, felsic igneous rock.

17. Which of the following is a characteristic of pegmatites?
    a. very large grain size
    b. gem minerals
    c. occurrence in granitic rocks
    d. all of these

18. Which of the following rocks has a glassy texture?
    a. granite
    b. obsidian
    c. tuff
    d. basalt

19. Which of the following rocks is pyroclastic?
    a. basalt
    b. tuff
    c. rhyolite
    d. andesite

20. Plutons form when
    a. magma solidifies in the subsurface.
    b. lava cools on the surface.
    c. a volcano erupts.
    d. none of these

21. Dikes are
    a. tabular and parallel to the surrounding rock layers .
    b. mushroom shaped.
    c. tabular and cross-cutting the surrounding rock layers.
    d. none of these

22. Batholiths are found in association with
    a. plateau basalts.
    b. oceanic islands.
    c. mountain ranges.
    d. the oceanic crust.

TRUE OR FALSE

___1.   Extrusive igneous rocks form from magma cooling in the subsurface.

___2.   Felsic igneous rocks are richer in silica than mafic igneous rocks.

___3.   The higher the viscosity of a lava, the faster it will flow.

___4.   A felsic magma has a lower viscosity than a mafic magma.

___5.   Most magma originates from the Earth's molten core.

___6.   As a magma cools, the first minerals to form are olivine and calcium plagioclase.

___7.   The discontinuous series consists of nonferromagnesian minerals.

___8.   The geothermal gradient is due to a decrease in temperature with depth.

___9.   Different minerals have different melting points.

___10. If a mafic magma of a given volume undergoes crystal settling, it will produce an equal volume of felsic magma.

___11. Assimilation refers to two magmas mixing together.

___12. The texture of an igneous rock is determined by the cooling history of the magma.

___13. Ultramafic lavas have been particularly abundant over the last 100 million years.

___14. The richer a magma is in silica, the lighter in color it will generally be.

___15. It would not be unusual to see olivine phenocrysts in a porphyritic basalt.

___16. Basalt is the most common extrusive igneous rock.

___17.   Abundant water vapor in the late stages of magma crystallization inhibits nucleation of crystals.

___18.   Obsidian is a common constituent of intrusive igneous rocks.

___19.   Granite is exposed on the surface only after uplift and erosion of the overlying rocks.

___20.   Laccoliths and sills have the same shape.

___21.   Batholiths are the largest of all plutons.

___22.   Granites can be both igneous and metamorphic in origin.

___23.   The presence of inclusions of country rock in a granite may indicate emplacement by stoping.

DRAWINGS AND FIGURES

1.   What was the cooling history of the rock shown in the figure to the right?

2.   What two types of plutons are shown in the figure below?

3. Using the diagram below, identify aphanitic and phaneritic rocks with the following compositions:
   a. 5% quartz, 38% potassium feldspar, 24% plagioclase feldspar, 15% biotite, 8% hornblende.
   b. 20% plagioclase, 8% hornblende, 56% pyroxene, 16% olivine
   c. 60% plagioclase, 1% biotite, 31% Hornblende, 8% pyroxene

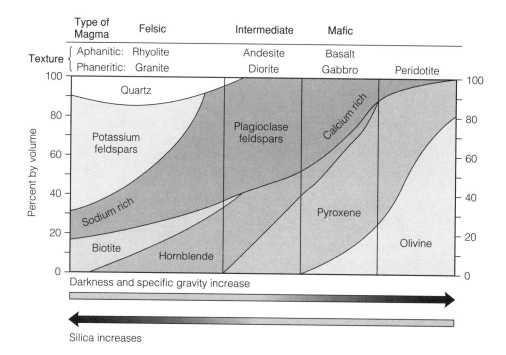

4. Sketch cross-sectional diagrams of the following plutons intruded into sedimentary (i.e. layered) country rock:
   a. stock
   b. sill
   c. laccolith
   d. dike

CHAPTER 4

VOLCANISM

CHAPTER OBJECTIVES

By the end of this chapter you should be able to:

1. Differentiate between active, dormant, and extinct volcanoes.
2. Be familiar with the different gasses, lavas, and pyroclastic materials released during volcanic eruptions.
3. Explain how a caldera forms.
4. List and differentiate between the major types of volcanoes.
5. Characterize the deposits of fissure eruptions and pyroclastic eruptions.
6. Describe the locations of several volcanic belts around the world and explain in a plate tectonic context why they occur where they do.

A USEFUL ANALOGY

Why does viscosity affect the type of eruption that a volcano will have? Think back to your childhood when you wouldn't have considered drinking milk with a straw without blowing bubbles in it. It was pretty easy to do, you didn't have to blow too hard. But what about a milkshake? That was tougher. You might have had to blow until red in the face. When you finally blew hard enough the milkshake likely exploded out of the cup. The reason is that the "thicker" milkshake required more pressure to get it out of the way of the rising gas (your breath). A magma that has a high viscosity similarly requires a high gas pressure to get it to move out of the way. Gas build until that pressure is finally reached. Then the rapid release of pressure may result in an explosive eruption.

KEY TERMS

After reading this chapter , you should be able to define the following terms:

volcanism	vog
active volcano	pahoehoe
dormant volcano	a a
extinct volcano	pressure ridge

spatter cones	composite cone
columnar joints	lahars
pillow lava	lava dome
ash	nuee ardente
lapilli	fissure eruption
volcanic bomb	basalt plateau
volcanic block	pyroclastic sheet deposit
volcano	welded tuff
crater	circum-Pacific belt
caldera	Mediterranean belt
shield volcano	subduction zone
cinder cone	hot spot

CHAPTER CONCEPT QUESTIONS

1. What are the differences between active, dormant and extinct volcanoes?

2. What are the common gasses associated with volcanism and what problems can they cause?

3. How does viscosity affect the style of a volcanic eruption and the characteristics of a lava flow?

4. Explain the origin of each of the following volcanic features:
   a. spatter cone   e. caldera
   b. columnar jointing   f. lava dome
   c. pressure ridge   g. welded tuff
   d. pillow lava   h. basalt plateau

5. What type eruptions and what composition of magma produces shield volcanoes? Why?

6. What types of deposits are found associated with composite cones? How does lava composition control the style of eruptions of composite cones.

7. How does plate tectonics control the distribution of volcanoes?

8. How can volcanism occur in the middle of a tectonic plate?

9. How do the Hawaiian Islands reveal the rate and direction of Pacific Plate movement?

COMPLETION QUESTIONS

1. A volcano which has not erupted in historic times but is expected to erupt again is said to be _____, whereas one which will probably never erupt again is _____.

2. By far, the most abundant gas associated with volcanism is _____.

3. When a crust forms on flowing lava it tends to crack and buckle forming _____.

4. A lava flow may contract as it cools forming polygonal cracks called _____.

5. Basalts erupting under water form a distinctive type of lava called _____.

6. Fine-grained pyroclastic material (that less than 2 mm in diameter) is called _____, and larger particles between 2 mm and 64 mm are _____.

7. Lava can be erupted from conical mountains called _____, or can be extruded from cracks in the Earth's crust called _____.

8. When a volcano erupts from its summit the lava comes out of a circular opening called a _____.

9. A _____ forms when the top of a volcanic mountain collapses into an empty magma chamber.

10. Big, broad volcanoes with gently sloping sides are called _____ volcanoes. These volcanoes erupt _____ lavas.

11. The most continuous eruption in history is the ongoing eruption of _____ .

12. Small volcanoes composed of pyroclastic debris are called _____ cones.

13. Large, steeply sloping conical volcanoes composed of alternating layers of lava and pyroclastic debris are called _____ cone volcanoes.

14. Large volcanic mudflows called _____, may make up a large portion of composite cones.

15. The Hawaiian Islands are dominated by _____ volcanoes.

16. The Cascade Range of the Pacific Northwest is dominated by _____ cones.

17. Extensive but relatively thin sheets of basaltic lava erupted from long fissures are called _____.

18. Sheet-like deposited similar to those described above, but made of material ejected by explosive eruptions are _____.

19. New oceanic crust is produced by volcanoes located along _____.

20. The circum-Pacific belt is a long belt of volcanoes located above _____ at convergent margins.

21. Volcanism at convergent plate boundaries is characterized by _____ volcanoes.

22. The Hawaiian Islands are located in the middle of the Pacific plate over a _____.

MULTIPLE CHOICE

1. If a volcano has erupted in historic times it is considered
    a . active.
    b. dormant.
    c. extinct.
    d. none of the above

2. The odorous haze known as vog is due which gas?
    a . water vapor
    b. hydrogen sulfide
    c. chlorine
    d. carbon monoxide

3. Lava which cools to form a pile of rough, jagged blocks is called
   a. pahoehoe.
   b. aa.
   c. nuee ardentes.
   d. none of the above

4. The more viscous a lava, the more
   a. slowly it flows.
   b. likely it is to erupt violently.
   c. silica it contains.
   d. all of the above

5. Ejected globs of molten lava which cool during flight are called
   a. pressure ridges.
   b. volcanic blocks.
   c. volcanic bombs.
   d. nuee ardentes.

6. When magma rises to the surface, gases are released because
   a. the pressure is reduced.
   b. the pressure is increased.
   c. the silica content changes.
   d. none of these

7. If tension cracks develop in a cooling lava, a _____ lava forms.
   a. blocky
   b. glassy
   c. columnar jointed
   d. none of these

8. Shield volcanoes typically have slopes
   a. from 2 to 10 degrees.
   b. of around 30 degrees at the summit to as low as 5 degrees at the base.
   c. as steep as pyroclastic material will rest (typically about 33 degrees).
   d. nearly 90 degrees.

9. Mt. Pinatubo and Mt. St. Helens are examples of which type of volcano?
   a. shield volcano
   b. cinder cone
   c. composite volcano
   d. fissure volcano

10. Pyroclastic debris forms
    a. when lava flows form a fissure.
    b. by a violent volcanic eruption.
    c. lava cools slowly.
    d. none of these

11. Crater Lake, Oregon is an example of
    a. a fissure.
    b. a volcanic crater.
    c. a caldera.
    d. none of the above

12. Lahars commonly make up large percentage of
    a. shield volcanoes.
    b. cinder cones.
    c. composite cones.
    d. fissures.

13. Eruptions of high volumes of low viscosity lava from long fissures result in
    a. basalt plateaus.
    b. pyroclastic sheets.
    c. lava domes.
    d. all of the above

14. The largest volcanoes on earth in terms of volume are
    a. shield volcanoes.
    b. cinder cones.
    c. composite cones.
    d. All of the above are about the same size.

15. The steepest type of volcano is
    a. shield volcanoes.
    b. cinder cones.
    c. composite cones.
    d. fissures.

16. A nuee ardente will most likely result in a
    a. pillow lava.
    b. welded tuff.
    c. basalt plateau.
    d. pressure ridge.

17. Basalt plateaus
    a. are due to fissure flows.
    b. are abundant in parts of Washington and Oregon.
    c. are very fluid lava flows.
    d. all of the above

18. Most of the Earth's major volcanoes occur
    a. along the circum-Mediterranean belt.
    b. over hot spots in the middle of plates.
    c. along the circum-Pacific belt.
    d. none of these

19. Submarine volcanic eruptions along mid-ocean ridges generally result in
    a. pillow basalts
    b. pyroclastic sheet deposits
    c. plateau basalts
    d. lahars

20. The Hawaiian Islands formed
    a. at a ridge crest.
    b. at a subduction zone.
    c. at a trench.
    d. none of these.

TRUE OR FALSE

___ 1.   A dormant volcano can become an active volcano.

___ 2.   Carbon dioxide is the most abundant volcanic gas.

___ 3.   Volcanic gasses alone cannot cause death if lava is not present.

___ 4.   In the years following a large eruption the Earth generally experiences an increase in average global temperature.

___ 5.   Pahoehoe can become aa as it flows away from the point of eruption.

___ 6.   Pillow lavas form when lave is extruded under water.

___ 7.   Ash seldom travels far from the point of eruption.

___ 8.   A caldera is relatively small compared to a crater.

___ 9.   Shield volcanoes are composed of about 99% lava flows.

___ 10.   Spatter cones are large steep cones composed of alternating pyroclastic materials, and lava flows.

___ 11.   Low viscosity lava flows tend to flow down predictable paths and so usually present minimum danger to human life.

___ 12.   Crater Lake should really be called Caldera Lake.

___ 13.   Cinder cones often form on the flanks of larger shield volcanoes or composite cones.

___ 14.   Cinder cones are generally not active as long as other types of volcanoes.

___ 15.   Composite volcanoes generally have steeper slopes near the base than at the summit.

___ 16.   Lava domes are usually mafic, but sometimes intermediate in composition.

___ 17.   Although common, nuee ardentes rarely constitute a danger to human life.

___ 18.   The Columbia Plateau basalts are due to lavas extruded from composite cone volcanoes.

___ 19.   When looking at the world as a whole, volcanic eruptions are quite random.

___ 20.   Traveling from the island of Hawaii to the island of Kauai, the Hawaiian islands become older.

___ 21.   The largest island in the Hawaiian chain, Hawaii, is the only one on which there are currently active volcanoes.

DRAWINGS AND FIGURES

1.   Draw a series of sketches illustrating the development of a caldera.

2.   Sketch a cross-sectional view of a shield volcano, a composite cone, and a cinder cone labeling lava flows, pyroclastic materials, the central vent, craters and calderas as appropriate.

3.   On the map of Hawaii below, label the youngest and oldest islands and indicate where active volcanism is taking place. Indicate with an arrow the direction of plate movement.

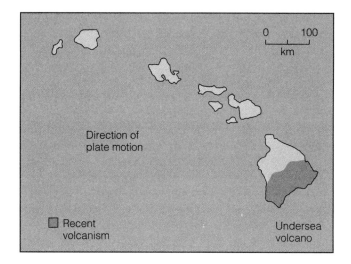

CHAPTER 5

WEATHERING, EROSION, AND SOIL

CHAPTER OBJECTIVES

By the end of this chapter you should be able to:

1. Differentiate between mechanical and chemical weathering.
2. List and explain six different mechanical weathering processes.
3. List and explain three different chemical weathering processes.
4. List and explain three different factors which control the rate of weathering.
5. List the essential components of soil and place them in the context of the soil profile.
6. List and explain five different factors which control rate and type of soil formed.
7. Describe several ways in which soil can be degraded.
8. List several mineral resources which are mined in the soil.

A USEFUL ANOLOGY

To see how surface area affects weathering rate dissolve a teaspoon of table salt in a glass or beaker. At the same time dissolve an equal volume of rock salt that you buy to thaw sidewalks in winter. If the latter is unavailable in your part of the country you may have a similar sized piece of halite in your lab. The table salt dissolves before coarser counterpart every time.

KEY TERMS

After reading this chapter , you should be familiar with the following terms:

weathering	talus
parent material	joints
erosion	frost heaving
transport	pressure release
differential weathering	sheet joints
mechanical weathering	exfoliation
chemical weathering	exfoliation domes
frost wedging	popping

thermal expansion
solution
carbonic acid
oxidation
hydrolysis
spheroidal weathering
regolith
soil
humus
loess
residual soil
transported soil
loess
soil horizons
topsoil
leaching

subsoil
zone of accumulation
pedalfer
pedocal
caliche
laterite
relief
bauxite
soil degradation
sheet erosion
rill erosion
salinization
residual concentration
gossan
oxidized ore
supergene enrichment

CHAPTER CONCEPT QUESTIONS

1.   Explain the difference between frost heaving and frost wedging.

2.   How are exfoliation domes formed?

3.   How does the excavation of deep underground mines cause sheet joints to form and what danger do they pose miners?

4.   How can forests fire act as a mechanical weathering agent?

5.   Describe a way in which the activities of organisms aid in mechanical weathering and a way in which they participate in chemical weathering.

6.   What happens at the molecular level when a mineral dissolves in water? When a mineral dissolves in a weak acid?

7.   How does clay form?

8.   Describe a climatic condition which would promote chemical weathering.

9.   How does Bowen's reaction series predict susceptibility of minerals to chemical weathering?

10.   Compare and contrast a soil found in the humid eastern U.S. with one found in the desert southwest and one found in equatorial Brazil.

11.   Explain why the richest aluminum ore deposits are a product of weathering.

12.   Explain the differences between, and give examples of, erosion, physical soil degradation, and chemical soil degradation. For each example think of a way that farming practices can mitigate damage.

COMPLETION QUESTIONS

1. The physical breakdown and chemical alteration of rocks and minerals on the surface of the Earth is called _____.

2. The removal of weathered material is called _____.

3. The process of two different rocks weathering at two different rates is called _____ weathering.

4. The two types of weathering are _____ and _____ weathering.

5. The repeated freezing and thawing of water in cracks of rock produces _____ wedging.

6. Exfoliation domes are due to a release in _____ because overlying rocks have been _____ away.

7. Chemical weathering can not occur without the presence of _____.

8. Common chemical weathering reactions include ions going into _____, iron bearing minerals being _____, and feldspars undergoing _____.

9. Hydrolysis reactions on _____ produces _____.

10. _____ size particles chemically weather fastest.

11. The early formed minerals of Bowen's reaction series weather _____ than the later formed minerals.

12. Naturally occurring surface materials that support plant life are called _____.

13. A soil derives its dark color from _____ matter.

14. Soils that form in place are called _____ soils.

15. In a soil profile, the _____ horizon contains mineral and organic matter, while the _____ horizon is a clay-rich layer.

16. The **A** horizon of a soil is called the zone of _____, and the **B** horizon is called the zone of _____.

17. The most important factor in determining rate and type of soil formed is _____.

18. If a soil has had most of its ions leached from the **A** horizon, but maintains a high organic content, it is classified as a _____.

19. A concentration of calcium carbonate in the **B** horizon is called _____ and is characteristic of _____ soils.

20. The concentration through leaching of aluminum hydroxides is characteristic of _____ soils and are a source of aluminum ore called _____.

21. The more time a rock body has been exposed to soil formation, the _____ the soil profile.

22. The two most significant factors in soil erosion are _____ and _____.

23. Weathering can lead to the formation of ore bodies by the selective removal of soluble substances to produce a _____ concentration.

24. Lead, zinc, copper, and nickel deposits may form where solutions containing these metals penetrate below the water table. These metals replace ions in the primary mineral deposits in a process called _____ enrichment.

MULTIPLE CHOICE

1. One expects talus cones to be most abundant in areas
   a. where abundant rainfall is available to dissolve limestone.
   b. in tropical areas were thick soil and abundant vegetation occurs.
   c. in high mountains that have many days of sub-freezing temperatures.
   d. none of the above

2. Exfoliation domes form from
   a. frost wedging.
   b. pressure release.
   c. hydrolysis.
   d. root wedging.

3. Plants may play an important role in
   a. mechanical weathering.
   b. chemical weathering.
   c. both **a** and **b**
   d. neither **a** nor **b**

4. Which of the following is NOT a mechanical weathering process?
   a. frost wedging
   b. pressure release
   c. hydrolysis
   d. root wedging

5. An important acid in the weathering of limestone is
   a. sulfuric acid.
   b. hydrochloric acid.
   c. acetic acid.
   d. carbonic acid.

6. Limestone and marble chemically weather by
   a. oxidation.
   b. hydrolysis.
   c. solution.
   d. all of these

7. Oxidation has the greatest effect on
   a. feldspars.
   b. ferromagnesian silicates.
   c. nonferromagnesian silicates.
   d. carbonates.

8. Feldspars weather to clay by
   a. solution.
   b. hydrolysis.
   c. oxidation.
   d. none of these

9. The asymmetry of the water molecule make it an important agent of
   a. exfoliation.
   b. frost heaving.
   c. solution.
   d. oxidation.

10. Which of the following rock types most commonly displays spheroidal weathering
    a. limestone.
    b. granite.
    c. rock salt.
    d. all equally

11. Chemical weathering is most intense in a
    a. hot, dry climate.
    b. cold, wet climate.
    c. hot, wet climate.
    d. cold, dry climate.

12. Which of the following is an example of regolith
    a. soil.
    b. sand dunes.
    c. volcanic ash.
    d. all of the above

13. The C horizon of a soil has more ____ than the horizons above it.
    a. clay minerals
    b. organic matter
    c. rock fragments of the parent rock
    d. none of these

14. The A horizon of a soil is the
    a. zone of accumulation.
    b. zone of leaching.
    c. zone of caliche.
    d. zone of calcification.

15. The O horizon of a soil is rich in
    a. clay minerals.
    b. organic matter.
    c. rock fragments of the parent rock.
    d. none of these

16. The factor most important in determining soil type and depth is
    a. parent material.
    b. plants.
    c. climate.
    d. they are all equal.

17. Pedocals are
    a. rich in calcium.
    b. found in the west.
    c. formed mostly in arid areas.
    d. all of the above

18. Caliche is an accumulation of
    a. calcium carbonate in the A horizon.
    b. calcium carbonate in the B horizon.
    c. aluminum in the B horizon.
    d. none of these

19. Laterite soils
    a. are red in color.
    b. are formed in tropical climates.
    c. often contain high concentrations of bauxite.
    d. all of the above

20. Which is not a process of soil degradation
    a. salinization.
    b. residual concentration.
    c. erosion.
    d. compaction.

21. Bauxite forms from
    a. the residual concentration of aluminum in the soil.
    b. oxidation reactions with descending metal rich fluids just above the water table.
    c. supergene enrichment of zinc and lead below the water table.
    d. all of these

TRUE OR FALSE

___ 1. Erosion is the physical and chemical breakdown of rocks.

___ 2. Frost heaving is due to the expansion of ice as water freezes in unconsolidated sediment and soil.

___ 3. Exfoliation may produce large rounded domes of rock in mountainous areas.

___ 4. The growth of salt minerals in the cracks of rock is an important chemical weathering process.

___ 5. Plants and animals can bring about mechanical weathering.

___ 6. Water is an essential ingredient in chemical weathering.

___ 7. The red color in many soils is the result of the oxidation of aluminum.

___ 8. Most clay minerals result from the hydrolysis of calcite.

___ 9. As the particle size decreases, the rate of chemical weathering increases.

___ 10.   Some rocks are more resistant to chemical weathering than others.

___ 11.   Quartz is the most resistant of the rock forming minerals.

___ 12.   Spheroidal weathering occurs because the flat faces of a rectangular block of rock weather faster than the corners.

___ 13.   Soil is a naturally occurring surface material that can support plant life.

___ 14.   The B horizon is a humus-rich zone.

___ 15.   Transported soils can form in stream sediments.

___ 16.   Caliche is a soil rich in iron and aluminum.

___ 17.   Bauxite is an ore of iron oxide.

___ 18.   Iron oxide can give soils a red color.

___ 19.   The longer an outcrop is exposed to soil formation, the thicker the resulting soil profile.

___ 20.   Rill erosion occurs when rain water flows down a slope in a small channel called a rill.

___ 21.   Over half of all soil erosion is caused by wind.

___ 22.   Such methods as crop rotation, terracing, contour plowing, and no-till planting have proven to be of almost no effect in helping to minimize soil degradation.

___ 23.   Supergene enrichment occurs below the water table.

DRAWINGS AND FIGURES

1. Identify and explain the origin of the weathering features shown in the photographs at right and below.

2.   Draw a typical soil profile. Fig. 5.18 provides an answer.

3.   In the figure below, label the (a) gossan, (b) the zone of supergene enrichment, (c) the zone where oxidation and leaching of sulfide minerals occurs, and the (d) zone where oxidized ores accumulate.

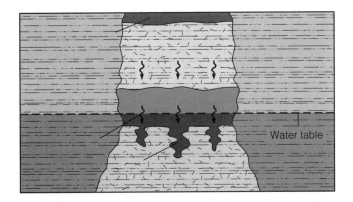

Water table

CHAPTER 6

SEDIMENT AND SEDIMENTARY ROCKS

CHAPTER OBJECTIVES

By the end of this chapter you should be able to:

1. Differentiate between detrital and chemical sediment.
2. Classify detrital sedimentary particles.
3. Describe what happens to detrital sedimentary particles during transport.
4. List several examples of sedimentary environments.
5. Describe the roles of compaction and cementation in of various sedimentary rocks.
6. Classify detrital sedimentary rocks and list most abundant constituents in each type.
7. Give the names and mineral compositions of the most abundant chemical and biochemical sedimentary rocks.
8. Explain what sedimentary facies are and how they become layers.
9. Describe and recognize several sedimentary structures, know how they form and what they say about the depositional environment.
10. Describe several ways in which fossils are preserved.
11. Be familiar with the important natural resources associate with sedimentary rocks.

A USEFUL ANALOGY

One of the harder concepts in geology for many people is understanding how a depositional environment like a beach, which has deposition in a very narrow zone, can become a layer that may cover hundreds of square miles. Related to this is the problem of getting facies that are lateral to one another to ultimately form vertically stacked layers. After all, there are no beaches hundreds of miles wide. The key is understanding the element of time and the concept of facies migration.

Imagine three workmen working in a house. They are in a big hurry to finish putting a carpet down and cover it with plastic (don't ask me why, its just an analogy). So they are all three working at once. The first workman is rolling out the carpet pad. Right on his heels is the chap rolling out the carpet itself on top of the pad. The third is rolling out the plastic on top of the carpet, right behind the second. At any point in time if you say "stop", each of the three will be in the act of rolling out his respective material. They will be doing this at only a single spot and they will be beside one another in space. This is like looking at a depositional system at any given point in time. Each workman is a facies depositing a different material. However, if you go back to where all three have been you will see a three part layering with a carpet pad on the bottom followed by a carpet and then the plastic...sedimentary layers!

KEY TERMS

After reading this chapter , you should be familiar with the following terms:

sedimentary rock	coal
sediment	anthracite
detrital sedimentary rock	bituminous
chemical sedimentary rock	sub-bituminous
gravel	lignite
sand	sedimentary facies
clay	marine transgression
rounding	marine regression
abrasion	sedimentary structures
depositional environment	strata
sorting	sedimentary bed
lithification	bedding plane
pore space	graded bedding
compaction	cross-bedding
cementation	turbidity current
clastic texture	paleocurrents
conglomerate	ripple marks
breccia	mud cracks
sandstone	fossil
arkose	trace fossil
mudrocks	petrification
siltstone	mold
mudstone	cast
claystone	concretion
shale	coke
biochemical sedimentary rocks	hydrocarbons
limestone	source rock
dolostone	reservoir rock
carbonate rocks	permeabiltiy
coquina	stratigraphic trap
chalk	structural trap
oolitic limestone	salt domes
evaporites	oil shale
rock salt	tar sands
rock gypsum	carnotite
chert	uraninite
flint	banded iron formations
jasper	

CHAPTER CONCEPT QUESTIONS

1.   Describe what happens to detrital particles during the transportation process.

2.   Compare and contrast the lithification of sand or gravel with that of mud.

3. A typical granite has about 60% feldspar and 15% quartz. However, sandstone derived from granite typically has over 60% quartz and 25% feldspar. Explain this reversal in abundance.

4. How do chemical sedimentary rocks differ in origin from detrital sedimentary rocks?

5. How do biochemical sedimentary rocks differ from other chemical sedimentary rocks?

6. Describe two ways in which limestone form. As an example contrast the origin of oolitic limestone with that of coquina.

7. Explain the origin of evaporites and give two examples?

8. How does chert form?

9. How does coal form?

10. What are sedimentary facies? How do they ultimately form layered rocks over time?

11. What are sedimentary structures, and how are they used? Give several examples of such structures and explain what information each provides about the origin of the rock in which you find them.

12. What is the difference between trace fossils and body fossils?

13. Describe several ways in which an organism can be fossilized.

14. Describe some of the tools that a geologist uses to identify the depositional environment of a sedimentary rock.

15. Explain the origin of oil and natural gas. How do they accumulate in economically viable deposits?

COMPLETE THE FOLLOWING TABLES:

DETRITAL SEDIMENTARY ROCKS			
Clast size	Sediment Name	Rock Name	
>2 mm		(rounded) (angular)	
1/16-2 mm			
1/256-1/16 mm			(mixture)
<1/256 mm			

CHEMICAL and BIOCHEMICAL SEDIMENTARY ROCKS	
Composition	Rock Name
Calcite	
Dolomite	
Gypsum	
Halite	
Quartz (microscopic)	
Carbon	

COMPLETION QUESTIONS

1. Sediment forms from the _____ of pre-existing rocks.

2. Rocks composed of the particles of disintegrated pre-existing rock are called the _____ sedimentary rocks.

3. Sedimentary particles are transported by wind, _____, and ice to a site of deposition.

4. While these sediments are being transported, they are reduced in size and made smoother by _____.

5. During transport, sediments are segregated by size in a process called _____.

6. Transportation ends when sediments are _____ in a depositional environment.

7. Once sediments are deposited, they are buried and converted to sedimentary rock in a process called _____.

8. In converting mud to shale, _____ alone is generally sufficient as a lithification process, but in converting sand to sandstone and gravel to conglomerate _____ is also important.

9. The fragments composing sedimentary rocks are called _____ and the texture exhibited by these rocks is, therefore, called a _____ texture.

10. Conglomerate and breccia are composed of clasts larger than _____ mm in diameter called _____ .

11. Compared to breccia, conglomerates display a greater degree of _____.

12. The most abundant mineral found as clasts in sandstone is _____.

13. The finest-grained detrital sedimentary rock is _____.

14. If mudstone or claystone displays _____ it is called shale.

15. A biochemical limestone composed entirely of broken shells is called _____.

16. _____ are small spherical grains of inorganically precipitated calcite.

17. Evaporite deposits form when seawater _____ and dissolved ions precipitate out.

18. Chert is a _____ sedimentary rock composed of _____.

19. _____ is composed of the compressed and altered remains of plants that were deposited in waters deficient in _____.

20. Coal grades from peat to _____, to _____, to _____.

21. During a marine _____, the sea advances on to a continent, and during a _____ the sea moves off of a continent.

22. Features that form as sediments are being deposited are called sedimentary _____.

23. Geologists use sedimentary structures to determine the environment of _____.

24. _____ can be used to determine the direction of flow of ancient currents.

25. A shell-shaped cavity caused by the dissolution of that shell is called a _____, and if this is filled-in to make a replica of the fossil, it is called a _____.

26. _____, _____ and _____ are energy resources that are found in sedimentary rocks.

27. Hydrocarbons form in a _____ rock, and accumulates in a _____ rock.

28. About _____ % of the world's proven reserves of petroleum are in the Persian Gulf region.

29. An important source of iron ore, banded iron formations are almost all _____ in age.

MULTIPLE CHOICE

1. Sedimentary rocks formed from particles of older rock broken up by weathering are called
   a. chemical sedimentary rocks.
   b. evaporites.
   c. detrital sedimentary rocks.
   d. biochemical sedimentary rocks.

2. Chemical sedimentary rocks from
   a. broken fragments of pre-existing rock.
   b. precipitation from solution.
   c. from wind blown sediments.
   d. all of the above

3. Sand size particles are
   a. 2-1/16 mm in diameter.
   b. greater than 2 mm in diameter.
   c. less than 1/15 mm in diameter.
   d. 1/16 to 1/256 mm in diameter.

4. Beaches, bays and lagoons are examples of
   a. marine environments.
   b. continental environments.
   c. transitional environments.
   d. none of these.

5. Sediments are transformed into sedimentary rocks by
   a. weathering and abrasion.
   b. compaction and cementation.
   c. sorting and abrasion.
   d. none of these

6. An arkose is
   a. sandstone composed of 90% quartz.
   b. a lithified gravel with angular clasts.
   c. a fissile mudrock.
   d. a sandstone with at least 25% feldspar.

7. Conglomerates are composed of
   a. angular grains larger than 2 mm in diameter.
   b. grains 2-1/16 mm in diameter.
   c. rounded grains larger than 2 mm in diameter.
   d. none of these

8. Shales are
   a. fissile.
   b. composed of particles <1/16 mm in diameter.
   c. the most abundant sedimentary rock.
   d. all of these

9. Limestone is composed of
   a. clay minerals.
   b. calcium carbonate.
   c. silica.
   d. none of these

10. Chalk is a type of
    a. limestone composed of microscopic shell fragments.
    b. limestone composed of oolites.
    c. sandstone composed of substantial amounts of feldspar.
    d. mudrock having fissility.

11. Chert is composed of
    a. halite.
    b. calcite.
    c. quartz.
    d. gypsum.

12. Coal forms from
    a. microscopic plankton.
    b. silica precipitated from seawater.
    c. calcite precipitated from seawater.
    d. none of these

13. The form of coal with the highest carbon contents is
    a. anthracite.
    b. peat.
    c. lignite.
    d. bituminous.

14. Evaporites are
    a. inorganically precipitated chemical sedimentary rocks.
    b. biochemical sedimentary rocks.
    c. detrital sedimentary rocks.
    d. none of the above

15. A transgressive sea
    a. produces no facies.
    b. moves off of a continent.
    c. advances into a continent.
    d. produces layers with shale underlying sandstone.

16. The most important use of sedimentary structures is to determine
    a. the environment of deposition.
    b. the age of a sedimentary rock.
    c. what economically important minerals might be present.
    d. what forms of life were present when the sediment was deposited.

17. Ancient current directions can be determined from
    a. mud cracks.
    b. current ripple marks and cross-bedding.
    c. graded bedding.
    d. trace fossils.

18. Fossils can be
    a. composed of the original bone and shell material.
    b. composed of bone and shells with their original composition replaced by a different mineral.
    c. composed of carbon films.
    d. all of these

19. A lens of permeable sandstone within an impermeable shale may be a
    a. source rock for hydrocarbons.
    b. a structural trap for hydrocarbons.
    c. a stratigraphic trap for hydrocarbons.
    d. all of the above

20. The richest uranium deposits are associated with
    a. evaporite deposits.
    b. banded iron formations.
    c. plant fossils in non-marine sedimentary rocks.
    d. fish fossils in deep marine sedimentary rocks.

TRUE OR FALSE

___1.   Detrital sediments form by precipitation from solution.

___2.   Sand size sediments are larger than 2 mm in diameter.

___3.   Sorting refers to the uniformity of the grain sizes present.

___4.   Compaction alone is generally enough to lithify mud into shale.

___5.   Breccias are composed of angular fragments larger than 2 mm in diameter.

___6.   Mudrocks may have both silt and clay-sized material.

___7.   Calcite is a common cementing mineral in sandstone.

___8.   Oolitic limestones are composed almost entirely of broken shell material.

___9.   Dolomite generally precipitates directly out of seawater.

___10.   The mineral found in rock salt is gypsum.

___11.   Flint is chert with organic impurities.

___12.   Coal forms as land plant debris accumulates in swampy water rich in oxygen.

___13.   Bituminous coal has a higher carbon content than lignite.

___14.   A facies is characterized by a distinctive set of physical, chemical, and biological attributes.

___15.   Deep sea shales are likely to be characterized by the presence of mud cracks.

___16.   Graded bedding shows an upward decrease in grain size.

___17.   Turbidity currents are flows of denser water flowing under less dense water.

___18.   When the original shell of a fossil dissolves it leaves a trace fossil.

___19.   Permeability refers to the capacity of a rock or sediment to transmit fluid.

___20.   Pure quartz sandstone is often mined for making glass.

___21.   Low permeability shale tends to make the best reservoir rocks.

___22.   The richest uranium ores in the United States are found in the Colorado Plateau.

___23.   Most banded iron formations formed over 2 billion years ago.

DRAWINGS AND FIGURES

1.   Identify the sedimentary structure in the photograph below. Indicate with an arrow the direction of flow recorded by this structure.

2.   Label the following depositional environments in the block diagram below.

submarine fan
continental shelf
organic reef
deep marine environment
fluvial environment
desert dunes
barrier island
alluvial fan

lake
tidal flat
beach
lagoon
glacial environment
shallow marine environment
delta

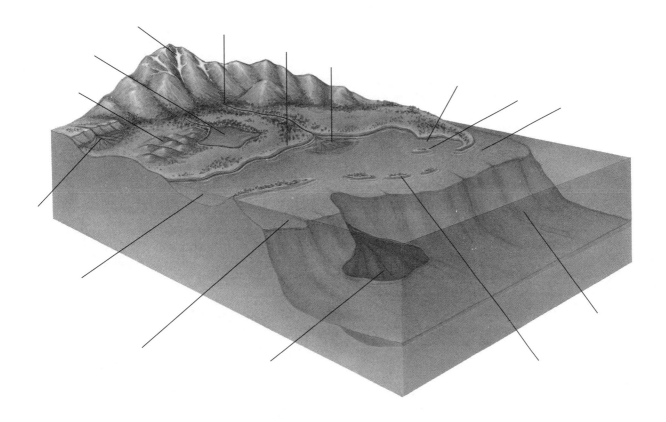

3. The block diagram below illustrates the distribution of facies adjacent to a shoreline. Sketch the vertical sequence of sedimentary rocks resulting from (a) a marine transgression and (b) a marine regression.

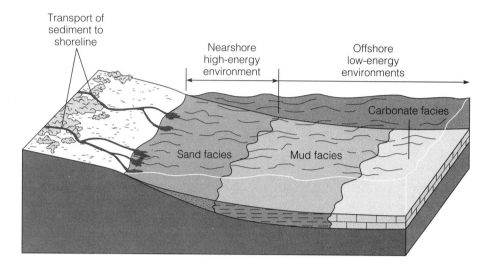

4. Identify the sedimentary structures in the following photographs.

CHAPTER 7

METAMORPHISM AND METAMORPHIC ROCKS

CHAPTER OBJECTIVES

By the end of this chapter you should be able to:

1. Explain why, and under what circumstances rocks metamorphose.
2. Locate the major shield areas of Earth.
3. Discuss how heat, pressure and fluid activity influence metamorphism.
4. Differentiate between contact, regional and dynamic metamorphism.
5. List the index minerals of metamorphic zones in order of increasing intensity of metamorphism.
6. Define and discuss the origin of foliation.
7. Classify metamorphic rocks listing important minerals, metamorphic grade, and parent rock of each.
8. Discuss the concepts of metamorphic zones and facies.
9. Relate types of metamorphism and metamorphic facies to plate tectonic setting.
10. List several economically important metamorphic rocks and minerals.

USEFUL ANALOGIES

It is easy to understand the difference between lithostatic and directed pressure if you realize that lithostatic pressure is a confining pressure and, therefore, much like two other types of confining pressure that we experience, hydrostatic pressure and atmospheric pressure. When you dive to the bottom of a swimming pool you do not feel the weight of the water column on your back. You do, nevertheless, experience an increase in pressure equally from all sides. This is may be manifested by a pain in your sinuses as the water tries to change the volume of your head much the same way it did the styrofoam cup in Fig. 7-4. When you rise and descend in an airplane you experience a similar change in pressure. As rocks are buried they experience a similar change in lithostatic pressure. You experience differential pressure whenever you feel the weight of an object on top of you. Pressure in this case is not equal in all directions.

By pretending to be a mineral you can see how differential pressure controls growth direction while confining pressure does not. Jump in a swimming pool and curl up in a fetal position. You are a mineral just starting to nucleate. To grow, stretch to the full length of your body. In a pool it doesn't matter what orientation you stretch. It is equally easy to grow in any direction. Now come out of the pool and curl up in a fetal position on the floor. Have a friend throw a heavy mattress on top of you.

You are now feeling the weight of the mattress and so you are experiencing differential pressure. Now try to grow by stretching to the full length of your body. While it is easy to grow in a horizontal orientation it is not so easy to grow vertically, against the directed pressure.

KEY TERMS

After reading this chapter , you should know the following terms.

metamorphic rock	phyllite
shield	schist
lithostatic pressure	schistosity
recrystallization	gneiss
differential pressure	amphibolite
asbestos	migmatite
fluid activity	marble
contact metamorphism	quartzite
aureoles	greenstone
hydrothermal alteration	hornfels
dynamic metamorphism	anthracite
mylonite	isograd
regional metamorphism	metamorphic zone
index minerals	metamorphic facies
foliated texture	graphite
slate	

CHAPTER CONCEPT QUESTIONS

1. What information about the earth does the study of metamorphic rocks provide?

2. What are the major changes that occur in a rock during metamorphism and what factors bring about those changes?

3. What is the effect of heat on metamorphism and what are the sources of the heat?

4. Explain the difference between lithostatic and differential pressure.

5. What is the effect of chemically active fluids on metamorphism and what are the sources of those fluids?

6. What are the three types of metamorphism and what are their major differences?

7. How does contact metamorphism take place? What are the major factors controlling the products of contact metamorphism?

8. What is dynamic metamorphism and under what circumstances does it occur?

9. How does regional metamorphism take place?

10. List the index minerals of regional metamorphism in order of increasing metamorphic intensity.

11. What is foliation and how is it produced? Why don't contact metamorphic rocks generally display foliation? Why are marbles and quartzites, even when created by regional metamorphism, nonfoliated?

12. What are metamorphic zones and metamorphic facies? How are they similar and how different?

13. List at least five economically important resources created by metamorphism?

COMPLETE THE FOLLOWING TABLES:

FOLIATED METAMORPHIC ROCKS			
Parent Rock(s)	Metamorphic Grade	Characteristics	Metamorphic Rock
mudrock, claystone volcanic ash	low	fine-grained, dull	
mudrock	low to medium	fine-grained, lustrous	
mudrock, impure carbonates, mafic igneous rock	low to high	distinct foliation, visible grains	
mudrock, sandstone, felsic igneous rock	high	light and dark minerals segregated into bands	
mafic igneous rock	medium to high	Hornblende dominates, dark, weak foliation.	

NONFOLIATED METAMORPHIC ROCKS			
Parent Rock(s)	Metamorphic Grade	Characteristics	Metamorphic Rock
limestone or dolostone	low to high	interlocking calcite and dolomite	
quartz sandstone	medium to high	interlocking quartz crystals	
mafic igneous rocks	low to high	fine grained, green color	
mudrock	low to medium	fine-grained, hard and dense	
coal	high	black, lustrous up to 98% carbon	

FOLIATED METAMORPHIC ROCKS			
Parent Rock(s)	Metamorphic Grade	Characteristics	Metamorphic Rock
	low	fine-grained, dull	Slate
	low to medium	fine-grained, lustrous	Phyllite
	low to high	distinct foliation, visible grains	Schist
	high	light and dark minerals segregated into bands	Gneiss
	medium to high	Hornblende dominates, dark, weak foliation.	Amphibolite

NONFOLIATED METAMORPHIC ROCKS			
Parent Rock(s)	Metamorphic Grade	Characteristics	Metamorphic Rock
	low to high	interlocking calcite and dolomite	Marble
	medium to high	interlocking quartz crystals	Quartzite
	low to high	fine grained, green color	Greenstone
	low to medium	fine-grained, hard and dense	Hornfels
	high	black, lustrous up to 98% carbon	Anthracite

COMPLETION QUESTIONS

1. Metamorphic rocks form from preexisting rocks when those rocks are subjected to conditions which change their _____ composition and/or _____.

2. Precambrian metamorphic rocks are exposed in vast areas of the continents called _____.

3. The agents of metamorphism are _____, _____, and chemically active _____.

4. Sources of heat include rising _____, and the _____ gradient.

5. _____ pressure is due to the weight of the overlying rocks and is applied equally in all directions.

6. In mountain belts, pressures from two directions often exceed pressures from other directions. These are called _____ pressures.

7. The presence of water _____ the speed of metamorphic chemical reaction.

8. Sources of water for metamorphic reactions include water trapped in the _____ of sedimentary rocks, volatile fluids from rising _____, and the release of water by the _____ of certain minerals.

9. The three major types of metamorphism are _____, _____, and _____ metamorphism.

10. _____ metamorphism is due to the "cooking" of country rock by magma.

11. Since temperature decreases away from a heat source, intrusions are commonly surrounded by concentric zones of different mineral assemblages called _____.

12. _____ metamorphism is generally restricted to fault zones.

13. The burial of large areas of rock into realms of high temperatures and pressures results in _____ metamorphism.

14. Foliated metamorphic rocks derive their texture from subjection to high _____ pressures.

15. _____ is a very fine-grained, low grade metamorphic rock, valuable because of its propensity to split along very flat "cleavage" planes.

16. _____ is composed of alternating bands of light colored quartz and feldspar and dark colored ferromagnesian minerals.

17. _____ appear to be mixtures of high grade metamorphic rocks and streaks of granitic igneous rocks.

18. Marbles are generally composed of the mineral _____ and/or dolomite and result from metamorphism of _____.

19. A very hard, strong metamorphic rock composed almost entirely of quartz is called _____.

20. Hornfels results from the _____ metamorphism of mudrocks.

21. The highest grade coal, _____, is actually a metamorphic rock.

22. A group of metamorphic rocks that are characterized by particular mineral assemblages formed under the same broad temperature pressure conditions is called a metamorphic _____.

23. Lines of equal metamorphic intensity plotted on a map are called _____.

MULTIPLE CHOICE

1. Metamorphic rocks form from
    a. igneous rocks only.
    b. sedimentary rocks only.
    c. igneous and sedimentary rocks only.
    d. igneous, sedimentary, and metamorphic rocks.

2. Metamorphism can occurs
   a. only after a preexisting rock melts and begins to cool.
   b. only in the solid state, before a rock melts.
   c. either a or b
   d. none of these

3. Metamorphism can cause a change in the
   a. texture of the pre-existing rock.
   b. mineral composition of the pre-existing rock.
   c. a and b
   d. none of these

4. With increasing depth
   a. temperature generally increases.
   b. lithostatic pressure increases.
   c. minerals tend to recrystallize.
   d. all of the above

5. The presence of water
   a. increases the rate of chemical reactions.
   b. reduces the rate of chemical reactions.
   c. has no effect on the rate of chemical reactions.
   d. prevents metamorphism from occurring.

6. Which factor is least important in contact metamorphism?
   a. heat
   b. high differential pressure
   c. presence of fluids
   d. all of the above are equally important

7. Contact metamorphism dominates
   a. in fault zones.
   b. at shallow depths around intrusions.
   c. in the cores of mountain belts.
   d. none of the above

8. Which of the following is most important in dynamic metamorphism?
   a. high lithostatic pressure
   b. high differential pressure
   c. presence of fluids
   d. All of the above are equally important.

9. Dynamic metamorphism dominates
   a. in fault zones.
   b. at shallow depths around intrusions.
   c. in the cores of mountain belts.
   d. none of the above

10. Regional metamorphism dominates
    a. in fault zones.
    b. at shallow depths around intrusions.
    c. in the cores of mountain belts.
    d. none of the above

11. Regional metamorphism usually involves an increase in
    a. heat.
    b. lithostatic pressure.
    c. differential pressure.
    d. all of the above

12. Which of the minerals listed below indicates the highest grade of metamorphism?
    a. chlorite
    b. garnet
    c. sillimanite
    d. biotite

13. Which of the following lists contains all foliated rocks?
    a. schist, gneiss, quartzite
    b. slate, schist, gneiss
    c. schist, gneiss, marble
    d. none of the above

14. The parent rock for marble is
    a. sandstone.
    b. mudrocks.
    c. limestone.
    d. any of the above

15. Which of the following is not metamorphic in origin?
    a. shale
    b. slate
    c. schist
    d. serpentine

16. Schist and gneiss form by
    a. contact metamorphism.
    b. regional metamorphism.
    c. dynamic metamorphism.
    d. hydrothermal alteration.

17. Metamorphic aureoles are associated with
    a. contact metamorphism.
    b. dynamic metamorphism.
    c. regional metamorphism.
    d. any of the above

18. Migmatites appear to be a mixture of
    a. granite and gneiss.
    b. basalt and slate.
    c. sandstone and schist.
    d. shale and phyllite.

19. Fault zones may be recognized by the presence of
    a. migmatites.
    b. isograds.
    c. aureoles.
    d. mylonites.

20. The most widespread regional metamorphism can be found
    a. at divergent margins.
    b. at convergent margins.
    c. at transform boundaries.
    d. over hot spots in the middle of plates.

TRUE OR FALSE

___1. Time is not an important factor in metamorphic reactions.

___2. Metamorphic and sedimentary rocks form the crystalline basement rocks found in the shields of the continents.

___3. An increase in temperature increases the speed of metamorphic reactions.

___4. Metamorphic grade decreases with increasing distance from the intrusion.

___5. Increases in lithostatic pressure generally impart a change in both volume and shape of a mineral.

___6. Contact metamorphism occurs when the rocks are in contact with a fault.

___7. Contact metamorphism can cause new minerals to form.

___8. Hot fluids is a by product of the metamorphism of many clay minerals.

___9. Hot, watery solutions from a crystallizing magma produce new metamorphic minerals by hydrothermal alteration of country rock.

___10. A mylonite is generally produced by hydrothermal alteration of country rock.

___11. The mineralogy of a metamorphic rock is a poor indicator of the degree or intensity of the metamorphism.

___12. Foliation results from an increase in lithostatic pressure.

___13. Hornfels is produced when shale is subjected to high temperatures but low differential pressure.

___14. Schist is characterized by having alternating bands of ferromagnesian and nonferromagnesian minerals.

___15. Phyllite generally indicate a higher grade of metamorphism than a gneiss.

___16. Index minerals are generally of little use in determining the intensity of metamorphism of pure marbles or quartzites.

___17. Mafic igneous rocks may metamorphose into amphibolite or greenstone.

___18. A metamorphic facies is named after its most characteristic rock or mineral.

___19. Sediments deposited in a trench are sheltered from metamorphism.

___20.   Divergent plate boundaries are likely places for contact metamorphism to occur.

FIGURES AND DIAGRAMS

1. Using the figure below, determine which mineral you would expect to find in (a) a rock subjected to intense regional metamorphism and (b) a rock subjected to contact metamorphism.

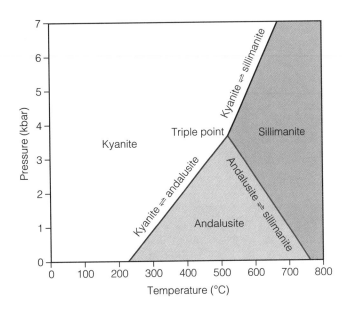

2. In the diagram below indicate where you would expect (a) contact (high temperature, low pressure), regional (high pressure, high temperature), and blueschist (high pressure, low temperature) metamorphism to occur. Be able to defend your choices.

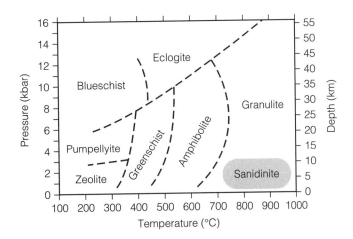

3.   Using the diagram below answer following questions.
 a.   Which facies would you expect to find in a schist metamorphosed from sediments deposited in a trench where high pressures but low temperatures exist?
 b.   As you progress away from a shallow (pressure = 1 kb) pluton which has a temperature of 700 degrees C, what progression of metamorphic facies would you expect to encounter?
 c.   As you progress away from the core of a mountain belt, where the most intense regional metamorphism occurred, toward the unmetamorphosed country rock, what progression of metamorphic facies would you expect to encounter?

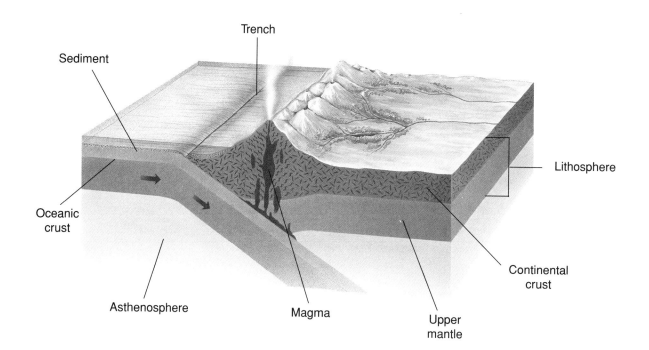

CHAPTER 8

GEOLOGIC TIME

CHAPTER OBJECTIVES

By the end of this chapter you should be able to:

1. Distinguish between relative dating and absolute dating.
2. Summarize the contributions to our view of geologic time of such notable scientists as James Hutton, Nicholas Steno, Lord Kelvin, Charles Lyell, Georges Louis de Buffon, and William "Strata" Smith.
3. List, define and apply those principles which are used in relative dating noting those that are attributed to Nicholas Steno.
4. Explain what an unconformity is, what it implies and classify the different types of unconformities.
5. Discuss the goals and techniques of correlation.
6. Recall the different parts of an atom.
7. Differentiate between the three types of radioactive decay and give examples of isotope parent-daughter pairs which exhibit each type.
8. Perform simple calculations of rock ages using half-lives.
9. Discuss the assumptions behind, and sources of uncertainty in radioactive dating.
10. Explain the principle behind fission track dating.
11. Explain technique and limitations of Carbon-14 dating.
12. Explain the principles behind dating using tree rings.
13. Reproduce the geologic time scale to the extent which your instructor requires.

USEFUL ANALOGIES

The vastness of geologic time is one of the hardest thing for mortals to grasp. Numbers like 4.5 billion are so unfamiliar to most people that to say that the Earth is that old is likely to elicit nothing but a blank stare. Over the years a number of techniques have been used to try to put the age of the Earth and the relative timing of events in some sort of perspective.

One popular way is to equate the age of the earth to a calendar year. The Earth was created at just an instant after midnight on January 1 and the present day is at midnight the following December 31. The oldest fossil life appears sometime before sunrise on March 25. However, abundant fossil life, with shells, that we all would recognize as fossils don't appear until between 6:00 and 6:30 PM

November 15! The first land plants sprout up about 5:30 PM on Nov. 26 and the first amphibians crawl onto land just before midnight on Dec. 1. The first reptiles show up around lunch time on Dec. 6 but the dinosaurs don't appear until 3:00 AM on Dec. 12. Mammals were a couple of days later, arriving on the scene about 345 on the 14th. Birds came along about 7:30 in the morning Dec. 20. The dinosaurs ruled the Earth until something killed them off mid-afternoon the day after Christmas. *Australopithecus* , the earliest hominid finally shows up about 3:30 in the afternoon on Dec. 31 and genus *Homo*, our very own genus, makes an appearance at 8:00 PM. *Homo sapiens* finally settles in about 12 seconds before midnight!

Another way is to set up a time line. You can make it 4.5 miles long so that the scale is 1 mile to a billion years. Life, then, becomes abundant over the last .57 miles of your line. Dinosaurs show up .25 miles from the end and die off about 77.5 feet from the end. *Australopithecus* arrives 5.25 feet from the end and *Homo sapiens* spans the last inch and a half. By the way your life is less than 3/10000 of an inch long. Feeling a bit insignificant?

Combining relative and absolute dating techniques is not limited to geology. We really do it all the time. Imagine going to meet a friend at a restaurant. On the way you notice that it starts to rain at 5:30. When you arrive your friend's car is already there. What can you say about how long he or she has been waiting? Answer: Look under the car. If it is wet you know the your friend arrived after the rain and has not been waiting all that long!

KEY TERMS

After reading this chapter you should know the following terms.

relative dating	assemblage range zone
absolute dating	isotope
principle of uniformitarianism	radioactive decay
principle of superposition	alpha decay
principle of original horizontality	beta decay
principle of lateral continuity	electron capture
principle of cross-cutting relationship	half-life
principle of inclusions	parent element
principle of fossil succession	daughter element
unconformity	mass spectrometer
disconformity	radon
angular unconformity	fission track dating
nonconformity	carbon 14 dating
correlation	tree-ring dating
key bed	geologic time scale
guide fossil	

CHAPTER CONCEPT QUESTIONS

1. What is the difference between relative time and absolute time?

2. Explain the principle of uniformitarianism.

3. How did Kelvin calculate the age of Earth? What were the problems with his theory?

4. Explain, with the help of diagrams if necessary, each of Steno's principles.

5. Explain, with the help of diagrams if necessary, the principle of cross-cutting relationships and the principle of inclusion.

6. Explain the principle of fossil succession. How has it proven valuable in correlation?

7. What is an unconformity? Summarize characteristics of the three major types.

8. List the desirable characteristics of an index fossil. How can fossils that are not index fossils be used in correlation?

9. Explain the differences between alpha decay, beta decay, and electron capture.

10. Why are sedimentary rocks not amenable to dating by radiometric techniques?

11. What are possible sources of error in dating rocks radiometrically? How can geologist mitigate some of the uncertainties?

12. Explain the principles behind dating by fission track method.

13. How does Carbon-14 form? Explain the principle behind its use in absolute dating.

14. How are tree rings used in absolute dating?

COMPLETION QUESTIONS

1. _____ dating placing geologic events in sequential order without regard to age in years.

2. _____ dating gives an age of a sample expressed in years before present.

3. Currently our best estimate for the age of Earth is _____ years.

4. The principle of uniformitarianism asserts that geologic _____ happening today have operated throughout geologic time.

5. The principle of superposition states that in undisturbed beds, the oldest beds are on the _____.

6. The principle of _____ states that sediments are deposited in horizontal layers.

7. If inclusions of one rock are found within another, the rock containing the inclusions are _____ than the rock from which the inclusions were derived.

8. If a fault cuts across a sandstone layer the sandstone is _____ in age that the fault.

9. Gaps in the geologic record of an area are called _____.

10. Unconformities with flat lying beds overlying tilted beds are called _____ unconformities.

11. Nonconformities occur when _____ beds overlie _____ or _____ rocks.

12. Disconformities are often difficult to identify and may depend on the study of the _____ in the sequence to recognize them.

13. Correlation is sometimes accomplished by using layers of distinctive composition called _____.

14. Guide fossils have a _____ geographic distribution and lived over a _____ time range.

15. Radioactive dates are based on the ratio of unstable _____ elements to their stable products called _____ elements.

16. In alpha decay a particle composed of two _____ and _____ neutrons are emitted from the nucleus.

17. In beta radiation a _____ in the nucleus changes to a _____ by emitting an _____.

18. The _____ of an element is the time it takes for half of the parent element to decay.

19. The amount of parent and daughter atoms present is determined by a sophisticated instrument called a _____ .

20. Radiometric dating is least useful in determining the ages of _____ rocks.

21. Carbon-14 dates are useful if the sample is less than _____ years old.

22. Carbon-14 decays to _____ by _____ decay and is not replenished once an organism dies.

23. Geologically recent events are sometimes datable by correlating the _____ of trees.

MULTIPLE CHOICE

1. Absolute dates are based on
   a. fossils.
   b. educated estimates.
   c. radioactive decay.
   d. cross-cutting relationships.

2. The methods that gave an acceptable date for the age of the Earth was
   a. the cooling rate of the Earth.
   b. the rate of sedimentation.
   c. the salinity of the oceans.
   d. none of the above

3. The principle of uniformitarianism required that the age of the Earth
   a. must be much older than previously thought.
   b. must be much younger than previously thought.
   c. could only be deciphered from studying the salinity of the ocean.
   d. could be deciphered from radioactive materials.

4. In undisturbed sedimentary deposits, the oldest beds
   a. are usually lava flows.
   b. are on the top.
   c. are on the bottom.
   d. could be b or c

5. Which of the following principles is NOT attributed to Steno
    a. original horizontality.
    b. uniformitarianism.
    c. lateral continuity.
    d. All of the above ARE attributed to Steno.

6. The Earth is approximately
    a. 4.6 billion years old.
    b. 4.6 million years old.
    c. 46 million years old.
    d. 6000 years old.

7. The principle of lateral continuity states that
    a. beds end abruptly.
    b. beds gradually thin and pinch out.
    c. the oldest beds are on the bottom.
    d. that beds are deposited as flat lying units.

8. The principle of fossil succession states that
    a. fossils of a particular organism occur at different times in different locations.
    b. the sequence of fossils is the same all over and, therefore, is predictable.
    c. fossils cannot be relied upon in correlation.
    d. none of the above

9. Unconformities represent
    a. a gap in the rock record.
    b. a period of erosion or nondeposition.
    c. neither a or b
    d. both a and b

10. A disconformity occurs where
    a. the beds are parallel above and below the unconformable contact.
    b. the beds above and below the unconformable contact are angles to one another.
    c. sedimentary rocks overlie massive crystalline rocks.
    d. all of the above

11. A nonconformity may closely resemble an intrusive contact, particularly in concordant plutons. Which principle may help in making the distinction?
    a. lateral continuity
    b. uniformitarianism
    c. inclusions
    d. fossil succession

12. Rocks in different areas may be correlated by having
    a. a distinctive key bed in both areas.
    b. the same fossil content.
    c. the same rock type.
    d. any of the above

13. A good characteristic for a guide fossil would be
    a. long life span as a species.
    b. a narrow geographic distribution.
    c. difficult identification.
    d. none of the above

14. An alpha particle is composed of
    a. two protons and two neutrons.
    b. a proton and an electron.
    c. a neutron and an electron.
    d. an electron.

15. The parent - daughter pair of U-235 and Pb-207 has a half life of 704 million years. If a rock had 1 million atoms of U-235 when it formed and now has 250,000 million atoms, how old is it?
    a. 704 million years
    b. 1.408 billion years
    c. 1 billion years
    d. 4.5 billion years

16. U-238 (Uranium) has an atomic # of 92 and atomic mass of 238. If it gives off 1 alpha particle and 2 beta particles it will become
    a. U-234 (Uranium: # of 92, mass of 234).
    b. Pb-206 (Lead: # of 82, mass of 206).
    c. Th-230 (Thorium: # of 90, mass of 230).
    d. Ra-226 (Radium: # of 88, mass of 226).

17. Of the list below, the rock type most easily dated radiometrically is
    a. granite.
    b. shale.
    c. limestone.
    d. All of the above equally easy.

18. Radioactive dates will be inaccurate if the sample has
    a. leakage of the daughter isotope.
    b. heating occurs during metamorphism.
    c. weathered.
    d. all of the above

19. Radioactive carbon dates are useful for the past
    a. 5000 years.
    b. 70,000 years.
    c. 250,000,000 years.
    d. 4.5 billion years.

20. The geologic time scale was developed using
    a. relative dating methods
    b. absolute dating methods
    c. a combination of a and b
    d. neither a nor b

TRUE OR FALSE

___1. Relative time tells us how long ago an event took place.

___2. The geologic time scale is based on rock sequences placed in chronological order.

___3. James Hutton's principle of uniformitarianism required that the earth must be older than anyone had previously thought.

___4.   The principle of superposition allows geologists to obtain an absolute date for a given rock layer.

___5.   If a fault cross-cuts a rock layer, then the fault is younger than the rock layer.

___6.   William Smith demonstrated that fossils succeed one another in time in a predictable order.

___7.   A disconformity often requires the use of fossils for its recognition.

___8.   A nonconformity exists when sedimentary beds underlie massive crystalline rocks.

___9.   An angular unconformity exists when the beds above and below the unconformity have an angular relationship.

___10. The process of establishing the time equivalency of rock units in different areas is called correlation.

___11. Correlation on the basis of rock type is only reliable in a fairly restricted area.

___12. Guide fossils have a restricted geographic range and short time of existence.

___13. After three half-lives a rock will only have 1/3 of the atoms of radioactive parent product that it originally had.

___14. Only under unusual circumstances can the radiometric age for sedimentary rocks be determined.

___15. Metamorphism can reset a rocks radioactive "clock".

___16. One advantage of radiometric dating is that it can be performed on weathered rock.

___17. One advantage of fission track dating is that it is one of the few ways to absolutely date the Pleistocene rocks.

___18. Carbon-14 dates have a short range of application because the half-life of carbon-14 is very short.

___19. Carbon-14 enters the bodies of living organisms in a constant ratio to other isotopes of carbon throughout that organisms life.

___20. Tree-ring dating has a range extending back to 14,000 years before the present.

DRAWINGS AND FIGURES

1.   The Grand Canyon is illustrated in the photograph below. What general statements can be made about the relative ages of these rocks and the relationship of the rocks in the foreground with those in the distance? What principles do you apply in making these statements?

2.   Sketch cross-sectional views of each of the three types of unconformities. Be sure to label rock type. Check your answers against Figures 8.9, 8.10, and 8.11.

3.   Using the basic principles of relative dating, list the rocks, unconformities and faults labeled in the block diagram below in order from oldest to youngest.

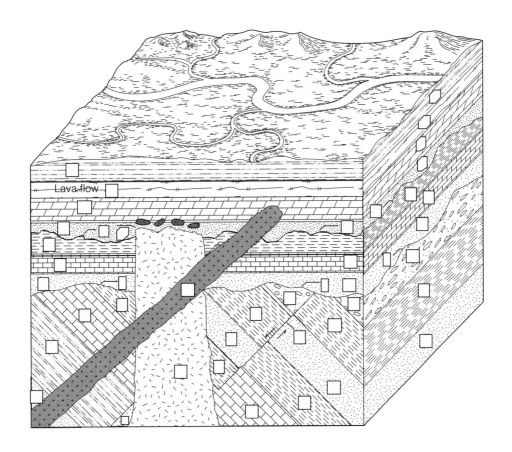

Lava flow

CHAPTER 9

EARTHQUAKES

CHAPTER OBJECTIVES

By the end of this chapter you should be able to:

1. Explain the elastic rebound theory for earthquakes.
2. Explain how a simplified modern seismograph works.
3. Differentiate between shallow, intermediate and deep focus earthquakes.
4. Describe where earthquakes occur and why they occur where they do.
5. Compare and contrast the various types of seismic waves including such aspects as relative velocities and direction of particle movement.
6. Determine the distance from a seismic station to an earthquake epicenter.
7. Determine the precise location of an earthquake epicenter.
8. Explain how earthquake intensity is determined.
9. Using the appropriate scale, determine the magnitude of an earthquake.
10. Produce a comprehensive list of damaging effects of earthquakes.
11. Define what is meant by "earthquake precursor" and give several examples.
12. Discuss the prospects for earthquake control.

USEFUL ANALOGIES

A Slinky® is a useful tool for demonstrating P and S waves. Stretch it out between two people. Let one person be the wave source. When the wave source gives a snap of the wrist, a wave will travel down the slinky, This wave will "vibrate" perpendicular to its direction of travel. After that, have the wave source give a sharp push parallel to the Slinky®. This push must be very sharp and quick. This time a wave will travel down the Slinky® but it will vibrate parallel in the same direction of travel being manifested as a compression and extension of the spring as it travels. Note which of the two migrates the fastest.

KEY TERMS

After reading this chapter , you should be familiar with the following terms.

earthquake	S-waves (secondary wave)
aftershock	shear waves
elastic rebound	elasticity
seismology	surface waves
seismograph	R-waves (Rayleigh waves)
seismogram	L-waves (Love waves)
seismic waves	time-distance graph
focus	intensity
epicenter	Modified Mercalli Intensity Scale
shallow focus	isoseismal lines
intermediate focus	magnitude
deep focus	Richter Magnitude Scale
Benioff zone	liquefaction
Circum-Pacific belt	tsunami
Mediterranean-Asiatic belt	seismic risk maps
body waves	precursors
surface waves	seismic gaps
P-waves (primary waves)	dilatancy

CHAPTER CONCEPT QUESTIONS

1. Explain the elastic rebound theory as a cause of earthquakes.

2. Explain, with sketches if desirable, how a seismograph works?

3. What is the difference between the epicenter of an earthquake and the focus?

4. Explain how earthquakes are distributed relative to plate boundaries? Where do deep and intermediate earthquakes occur and why?

5. What is the difference between body and surface waves?

6. List the four types of seismic waves in order of arrival at a seismic station (i.e. decreasing velocity). For each type describe the "vibration direction" or direction of material movement relative to the direction of wave travel.

7. How is the distance from a seismic station to an earthquake epicenter determined?

8. How do seismologists precisely locate an earthquake epicenter?

9. How is earthquake intensity determined? What are these values used for?

10. How is earthquake magnitude determined? How does it relate to energy released during an earthquake?

11. How does the ground material upon which a structure is built influence the extent to which it might be affected by ground shaking during and earthquake?

12. What sort of precautions should developers in earthquake-prone areas take to mitigate the potential damages from earthquakes?

13. Besides shaking of the ground, what are some potential destructive effects of earthquakes?

14. Describe four earthquake precursors.

15. What are seismic gaps and how are they used in earthquake prediction?

16. Describe a possible technique for controlling earthquakes that was suggested by experiments in old Colorado oil fields?

COMPLETION QUESTIONS

1. An earthquake is a shaking of the earth which occurs when _____ stored in rocks undergoing _____ deformation is released by displacement of those rocks along faults.

2. The above explanation for the origin of earthquakes is called the _____ theory.

3. The study of earthquakes is called _____.

4. The instrument used to record seismic waves is a _____ and the record made by this instrument is called a _____.

5. The point of origin of an earthquake in the subsurface is termed the _____, and the point on the surface directly above is called the _____.

6. Earthquakes classified as "shallow" have foci located less than _____ km below the surface. Those classified as "deep" have foci located over _____ km below the surface.

7. Focal depths increase along a dipping zone beneath convergent plate margins called _____ zones.

8. 80% of all earthquakes occur in the _____ belt.

9. The _____ belt accounts for about 15% of all earthquakes, the remaining 5% occur in the _____ of plates and along ocean ridges where the plates are _____.

10. The number of earthquakes recorded each year by seismographs is _____.

11. Seismic waves that travel through the Earth are called _____ waves, and those that travel along the ground surface are called _____ waves.

12. Two types of body waves exist, _____ waves _____ waves.

13. The fastest moving waves, _____ , cause earth material to move back and forth _____ to the direction of wave motion.

14. _____ are the second type of waves to arrive at a seismic station. They cause earth materials to move up and down _____ to the direction of travel.

15. The two most important types of surface waves are _____ and _____ waves.

16. By measuring the difference between the arrival times of _____ and _____ waves, the distance to, and the location of an earthquake can be determined.

17. Earthquake _____ is based on building damage and personal observations and is measured according to the _____ scale.

18. Earthquake _____ is based on energy released during an earthquake and is measured according to the _____ scale.

19. A seismogram showing the record of an earthquake of magnitude 5 has a peak with ____ times the amplitude of one recorded during an earthquake of magnitude 3.

20. In an earthquake, ground shaking is the most severe in areas of loose, unconsolidated and water-logged material when the ground starts behaving as a fluid. This process is called _____.

21. As a result of an earthquake, coastal regions could be devastated by a _____.

22. Currently geologists are trying to identify _____ events to predict earthquakes, but no consistently reliable method currently exists.

MULTIPLE CHOICE

1. When a material returns to its original shape after the release of stress it is said to be
   a. ductile.
   b. brittle.
   c. elastic.
   d. none of these

2. The instrument that detects and records seismic waves is called a
   a. seismograph.
   b. seismologist.
   c. seismogram.
   d. none of these

3. Approximately 90% of all earthquakes are _____ focus.
   a. shallow
   b. intermediate
   c. deep
   d. out of

4. The point on the surface, directly above the subsurface location where an earthquake originates is called the
   a. focus.
   b. hypocenter.
   c. epicenter.
   d. all of the above

5. The most destructive earthquakes have
   a. shallow focus.
   b. intermediate focus.
   c. deep focus.
   d. All of the above are equally destructive.

6. _____ focus earthquakes occur at convergent margins.
   a. Shallow
   b. Intermediate
   c. Deep
   d. all of the above

7. Most of the world's earthquakes occur
   a. in the Circum-Pacific belt.
   b. in the Mediterranean-Asiatic belt.
   c. in the interior of plates.
   d. They are equally distributed around the earth.

8. Earthquakes occur where plates
   a. slide past each other.
   b. converge.
   c. diverge.
   d. all of the above

9. S-waves will travel through
   a. liquids.
   b. solids.
   c. air.
   d. all of the above

10. The ranking of seismic waves from fastest to slowest is
    a. L,S,P.
    b. P,S,L.
    c. P,L,S.
    d S,P,L.

11. Factors which control seismic wave velocity include
    a. density and dilatancy.
    b. elasticity and resistivity.
    c. density and elasticity.
    d. pressure and dilatancy.

12. The intensity scale is
    a. based on building damage and personal observation.
    b. based on the amplitude of seismic waves.
    c. based on the magnitude of seismic waves.
    d. another name for the magnitude scale.

13. The intensity of an earthquake is controlled
    a. by the size and duration of the earthquake, the depth of the focus, and distance from the epicenter.
    b. by the local geology.
    c. the population density and type of building construction.
    d. all of the above

14. A building will sustain the least amount of damage if it is built on
    a. artificial fill.
    b. water saturated sediments.
    c. unconsolidated sediments.
    d. bedrock.

15. An earthquake's magnitude takes into account
    a. the amount of building damage.
    b. the amount of energy released by the earthquake.
    c. the number of people displaced.
    d. all of the above

16. The Richer scale uses the _____ peak on a seismogram and so represents an instant in time. This means that the total energy of the earthquake is _____.
    a. highest / overestimated
    b. highest / underestimated
    c. lowest / overestimated
    d. highest / underestimated

17. An earthquake of magnitude 4 releases _____ times the energy as one of magnitude 2.
    a. 2
    b. 60
    c. 100
    d. 900

18. Tsunamis are large destructive waves that are due to
    a. earthquakes.
    b. submarine landslides.
    c. volcanic eruptions.
    d. all of the above

19. By far most of the damage caused by the great 1906 San Francisco earthquake resulted from.
    a. ground shaking
    b. ground rupturing
    c. fire
    d. tsunami

20. Events that geologists monitor to try and predict earthquakes include
    a. seismic gaps.
    b. changes in elevation and tilting of the ground.
    c. fluctuations of water levels in water wells, variations in the Earth's magnetic and electric resistance.
    d. all of the above

TRUE OR FALSE

___1.   A seismogram is the instrument used to record seismic waves.

___2.   The epicenter of an earthquake is always located directly over the focus.

___3.   Earthquakes along a transform plate boundary generally have a deep focus.

___4.   The San Andreas fault is part of the circum-Pacific belt where most of the world's earthquakes occur.

___5.   Most earthquakes have a deep focus.

___6.   Earthquakes occurring in the interior of a plate, far from boundaries are not large or dangerous.

___7.   P-waves will not travel through liquids.

___8.   P-waves are the most damaging type of earthquake waves to building foundations.

___9.   In S-waves, earth materials move up and down, perpendicular to the direction of movement of the wave.

___10.   The distance from a seismograph to the epicenter of an earthquake is determined by using the difference in arrival times between P and S waves.

___11.   The precise location of an earthquake can be determined with as few as two seismograph stations.

___12.   A well made, accurate seismograph is fundamental to determining earthquake intensity.

___13.   Insurance companies are more concerned with magnitude than intensity.

___14.   An isoseismal line is a line of equal magnitude.

___15.   Analysis of earthquake damage has repeatedly shown that ground shaking is most severe in areas of bedrock.

___16.   On a worldwide basis, there are more small earthquakes each year than large ones.

___17.   An earthquake of magnitude 6 releases twice as much energy as a magnitude 3 earthquake.

___18.   An earthquake of intensity X will register 10 times the amplitude on a seismogram as one of intensity IX.

___19.   Coastal areas thousands of miles from an earthquake can be affected by a tsunami.

___20.   Currently, geologists can not consistently and accurately predict earthquakes.

DRAWINGS AND FIGURES

1. On the figure below, highlight the area in which you would expect find foci (a) shallow, (b) intermediate, and (c) deep focus earthquakes.

2. Using the graph to the right, determine the distance between your seismograph station and the epicenter of an earthquake which registers 5 minutes between the arrival of P and S waves on your seismograph.

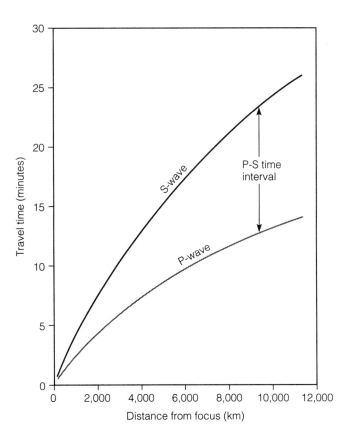

3. You are 20 km away from an earthquake which shows an amplitude of 100 mm on your seismogram. Using the graph below, determine its magnitude.

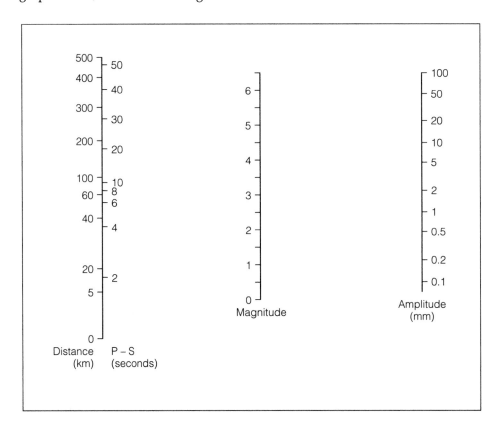

CHAPTER 10

THE INTERIOR OF THE EARTH

CHAPTER OBJECTIVES

By the end of this chapter you should be able to:

1. Describe changes in velocity and direction of seismic waves as they travel through the Earth and the reasons for these changes.
2. Explain why we think that we have a solid and liquid core.
3. Characterize each layer of the Earth in terms of composition, density, percentage of total Earth volume, and percentage of total Earth mass.
4. Differentiate between the distinct zones recognized in the mantle and characterize each.
5. Summarize the information that seismic tomography has contributed to our model of the mantle.
6. Account for the variations in thickness of the crust.
7. Recall the following values: (a) the geothermal gradient in the crust and (b) the in the mantle; the estimated temperature at the (c) base of the crust, (d) the core-mantle boundary, and (e) the center of the core.
8. List the areas of higher and lower than average heat flow.
9. Explain why the gravitational attraction between a suspended mass and the earth varies from place to place on the Earth.
10. Explain what isostasy is and why it occurs.
11. Differentiate between inclination and declination.
12. Explain what magnetic anomalies are and why they occur.
13. Explain what magnetic reversals are and what record they leave in the rocks.

KEY TERMS

After reading this chapter you should be familiar with the following terms.

crust
mantle
core
wave rays
refraction
reflection

discontinuity
P-wave shadow zone
S-wave shadow zone
Mohorovicic discontinuity (Moho)
low velocity zone
lithosphere

asthenosphere
transition zone
peridotite
seismic tomography
continental crust
oceanic crust
geothermal gradient
heat flow
gravimeter
gravity anomalies
positive gravity anomaly
negative gravity anomaly
principle of isostasy
isostatic rebound

magnetic field
dipolar
Curie point
magnetite
magnetic inclination
magnetic declination
magnetic anomalies
magnetometer
positive magnetic anomaly
negative magnetic anomaly
paleomagnetism
magnetic reversals
normal polarity
reverse polarity

CHAPTER CONCEPT QUESTIONS

1.   Why do seismic waves refract and reflect?

2.   How is the depth to the various boundaries in the earth (e.g. crust-mantle boundary) determined using seismic waves?

3.   What is the evidence that the outer core is liquid? What is the evidence that there is a solid inner core?

4.   What is the basis for current theories on the composition of the Earth's core?

5.   What is the Moho and how was it discovered?

6.   What is the seismic evidence for the asthenosphere?

7.   What is believed to be the cause of the discontinuities of the transition zone?

8.   What is seismic tomography? What has it contributed to our models of the mantle and core?

9.   Compare and contrast oceanic and continental crust in terms of composition, thickness and density.

10.   Why do geophysicists think that the geothermal gradient in the mantle is much less than in the crust?

11.   What factors determine the force of gravity? Why is the gravitational attraction between the Earth and moon so much more than between Earth and a small space craft an equal distance away? Why is the gravitational pull of the Earth on the space craft higher when it is in our atmosphere than when it is beyond our atmosphere? How does the gravitational attraction of the Earth on a gravimeter over an iron ore body compare to the attraction over a salt dome and why?

12.   Why does a mountain range rise vertically as it is eroded?

13.   Why don't geologists think that the Earth's magnetic field is due to a large body of magnetic material, such as magnetite, below the surface? What alternative origin for the magnetic field has been favored?

14.   What is the difference between inclination and declination?

15. What is the evidence that the Earth has reversed magnetic polarity from time to time?

COMPLETE THE FOLLOWING TABLE:

LAYER	COMPOSITION	DENSITY	% TOTAL VOLUME	% TOTAL MASS
Inner Core				
Outer Core				
Mantle				
Cont. Crust				
Oceanic Crust				

COMPLETION QUESTIONS

1. The Earth is composed of concentric layers of varying thickness differing in _____ and _____.

2. The outermost layer, called the _____, is divided into the thinner _____ crust and the thicker _____.

3. In very broad terms, the ocean crust is _____ and the continental crust is _____ in composition.

4. Below the crust is the iron-magnesium rich _____ which comprises more than 80 percent of the Earth's _____.

5. The mantle rests on an outer liquid _____ composed mostly of _____ but with 12% sulfur, silicon, oxygen, nickel, and potassium.

6. Below the outer liquid core is the inner core which is composed of _____ and _____.

7. Knowledge of the internal structure of the Earth is derived primarily from the behavior and travel times of _____.

8. As seismic waves move through the Earth, they refract when entering material of different _____ and _____.

9. Refraction involves a change in a wave's _____ and _____.

10. When a seismic wave reaches a boundary separating materials of significantly different _____ and _____, some of the energy is _____ back to the surface.

11. Marked changes in velocity of seismic waves indicate a _____, an example of which is the boundary between the crust and the mantle called the _____.

12. The nature of the core was postulated from the observations of _____ zones of P- and S-waves.

13. P-wave shadow zones extend between ____ and ____ degrees in either direction measured from an earthquake's _____.

14. A discontinuity within the mantle, at a depth of about 400 km, has been attributed to changes in _____.

15. Through the crust the geothermal gradient averages _____ per kilometer deeper in the mantle seems to decrease to as low as _____ per kilometer.

16. A sensitive instrument called a _____ detects variations in gravitational attraction of the earth at different locations.

17. If the gravitational attraction is higher or lower than expected for a given elevation, then a gravity _____ is said to exist.

18. The principle of _____ refers to Earth's crust floating on the _____.

19. As a result of isostatic equilibrium, the addition of weight on the Earth's crust causes it to _____, while the removal of weight from the Earth's surface cause the crust to _____.

20. The Earth's magnetic field appears to be due to electrical currents in the _____.

21. When rocks are heated to their _____ they lose their magnetism.

22. With time, the strength of the magnetic field weakens, and the _____ of the field can reverse.

23. Deviation of the magnetic field from the horizontal is called magnetic _____ and the angle between geographic north and magnetic north is called the _____.

24. Variations in the strength of the magnetic field are called magnetic _____.

25. _____ is the remnant magnetism in ancient rocks that records the direction and strength of the magnetic field at the time of their formation.

26. Rocks that have a record of magnetism the same as today's are described as having _____ polarity, whereas rocks with the opposite magnetism have _____ polarity.

MULTIPLE CHOICE

1. Generally, the velocity of seismic waves traveling through the mantle _____ with depth.
   a. increases
   b. decreases
   c. fluctuates randomly
   d. does not change

2. At the boundary between the mantle and the outer core P-waves suddenly
   a. increase in velocity.
   b. decrease in velocity.
   c. disappear entirely.
   d. are all reflected back to the surface.

3. Geologists know that the outer core is in the liquid state because
   a. P-waves will not travel through this region.
   b. L-waves will not travel through this region.
   c. S-waves will not travel through this region.
   d. all of the above

4. The mantle is composed of material rich in
   a. iron and nickel.
   b. iron and magnesium.
   c. iron and sulfur.
   d. silicon and aluminum.

5. The Moho is the boundary between
   a. the mantle and the core.
   b. the crust and the core.
   c. the upper mantle and the lower mantle.
   d. none of these

6. The low velocity zone roughly corresponds to the
   a. lithosphere.
   b. asthenosphere.
   c. outer core.
   d. inner core.

7. In broad, general terms, the
   a. oceanic crust is granitic, and the continental crust is basaltic.
   b. oceanic crust is ultramafic, and the continents are granitic.
   c. oceanic crust is basaltic, and the continents are granitic.
   d. none of the above

8. The continent crust is _____ the oceanic crust.
   a. thicker and less dense than
   b. thinner and less dense than
   c. thicker and more dense than
   d. equal in thickness and density to

9. The "hot" and "cold" areas revealed by seismic tomography are thought to reflect _____ in the mantle.
   a. changing crystal structure.
   b. discontinuities.
   c. shadow zones.
   d. convection cells.

10. Most of the heat lost by the Earth (about 70%) is lost through the
    a. high mountains of the continents.
    b. the flat areas of continents.
    c. the sea floor.
    d. the three deepest ocean trenches.

11. The force of gravity varies with the
    a. latitude and longitude.
    b. elevation and latitude.
    c. longitude and elevation.
    d. none of these

12. Force of gravity between two bodies
    a. increases as the masses decrease.
    b. increases as the masses move farther apart.
    c. increase as centrifugal force increases.
    d. none of the above

13. Over a buried mass of iron ore, gravity measurements would
    a. show a negative gravity anomaly.
    b. be lower than the adjacent area.
    c. be equal to that of the adjacent areas.
    d. be higher than that of the adjacent areas.

14. An example of isostatic rebound is
    a. the rise of continental crust as glaciers melt.
    b. the rise of mountains even as they erode.
    c. the maintenance of 10% of an iceberg above the surface of the water no matter how much melts.
    d. all of the above

15. Gravimeter readings across mountains show
    a. a high gravity anomaly.
    b. a low gravity anomaly.
    c. a low anomaly followed by a high anomaly.
    d. no gravity anomaly.

16. Which of the following would be expected to yield a negative gravity anomaly?
    a. an ore body
    b. a mountain range
    c. a salt dome
    d. all of the above

17. Deviations from the normal magnetic field strength are called
    a. magnetic inclinations.
    b. magnetic declinations.
    c. magnetic anomalies.
    d. none of these

18. Magnetic reversals occur when the
    a. polarity reverses.
    b. magnetic declination increases.
    c. magnetic inclination decreases.
    d. none of these

19. The Earth's magnetic field is due to
    a. changes in the rotation of the Earth
    b. a highly magnetic core.
    c. electrical currents in the outer liquid core.
    d. electrical currents in the mantle.

20. Over the last century the Earth's magnetic field has
    a. decreased in strength.
    b. increased in strength.
    c. stayed the same.
    d. reversed polarity.

TRUE OR FALSE

___1.   The Earth's average density is considerably greater than that of the surface rocks.

___2.   Wave velocity increases with increasing elasticity.

___3.   The S-wave shadow zone is due to the refraction of waves as they travel through the inner core.

___4.   The mantle is composed mostly of gabbro.

___5.   The Moho is the boundary between the outer and inner core.

___6.   The transition zone roughly corresponds to the asthenosphere.

___7.   Once below the asthenosphere, the mantle is homogeneous.

___8.   The continental crust is being stretched and thinned in areas like the East African Rift Valley and the Basin and Range Province of the U.S.

___9.   Oceanic crust is thickest at the spreading ridges.

___10.   Most of Earth's internal heat is generated by radioactive decay in the core.

___11.   The temperature in the Earth's core is thought to be close to the temperature of the surface of the sun.

___12.   Heat flow is highest at the spreading ridges.

___13.   The force of gravity in a given spot can be affected by the density of the material beneath the ground.

___14.   A huge positive gravity anomaly exists under high mountain ranges such as the Himalayas.

___15.   Since the last ice age much of the northern hemisphere has been subsiding.

___16.   The Earth is an example of a dipolar magnetic field.

___17.   Heating magnetic minerals to their Curie Point improves or enhances their magnetism.

___18.   The 11.5 degrees separating the magnetic north pole from the geographic north pole is called the magnetic inclination.

___19.   Geologic materials that add to the measured magnetic field produce positive magnetic anomalies.

___20.   During a period of magnetic reversal, the north arrow on a compass needle would point south.

___21.   Basaltic lava flows a are very useful rock in preserving paleomagnetism.

DRAWINGS AND FIGURES

1.   In the figure below label the layers of the Earth. In the inset in the upper right label the lithosphere, asthenosphere, upper mantle and the two types of crust. Check your answer against Figure 10-2 in your textbook.

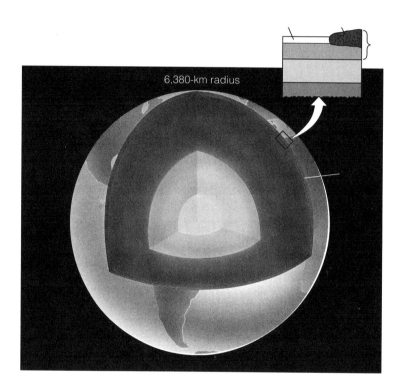

2. On the adjacent graph of P and S wave velocity vs. depth, indicate the location of (a) the crust-mantle boundary, (b) boundary between the outer core and mantle, and (c) the low velocity zone.

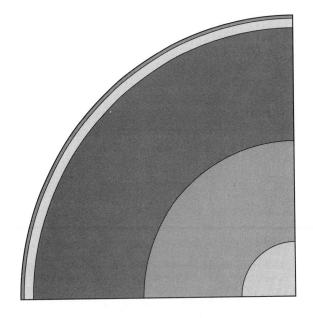

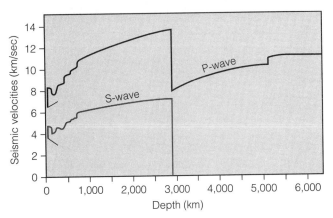

3. The adjacent figure shows the various paths of P waves through the earth from and earthquake focus. Label the P and S wave shadow zones.

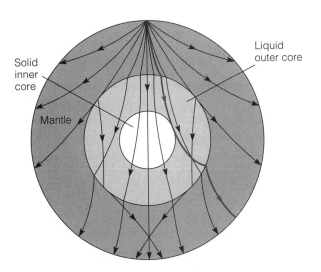

4.   In the three diagrams below, A and B are stations at which magnetic and gravity readings are made. At each station indicate positive and negative gravity and magnetic anomalies.

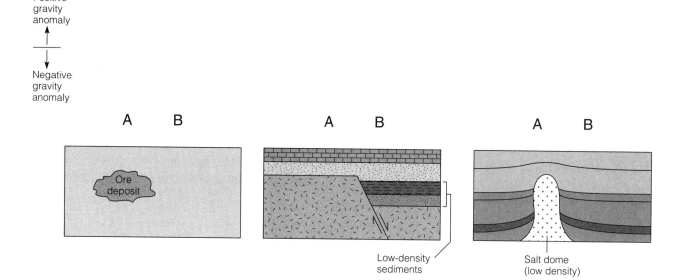

CHAPTER 11

THE SEA FLOOR

CHAPTER OBJECTIVES

By the end of this chapter you should be able to:

1. Compile a list of tools used in modern oceanographic research.
2. Describe the parts of a continental margin and classify continental margins by type.
3. Discuss the characteristics and effects of turbidity currents.
4. Describe the major features of the deep sea floor and give examples.
5. Characterize the types, and give the sources of, the various types of deep sea sediment.
6. Recount the evolution of a reef.
7. Describe a vertical profile through the sea floor and tell what evidence has lead to the development of that model.
8. List several economically important mineral resources which are found in the oceans.

KEY TERMS

After reading this chapter , you should be familiar with the following terms.

outgassing
H.M.S. *Challenger*
Deep Sea Drilling Project
Glomar Challenger
Alvin
echo sounder
seismic profiling
continental margins
continental shelf
continental slope
shelf-slope break
continental rise
turbidity currents
graded bedding
submarine fans

submarine canyons
active continental margin
passive continental margin
abyssal plains
oceanic trenches
oceanic ridge
seamounts
guyots
abyssal hills
aseismic ridges
microcontinents
manganese nodules
pelagic
pelagic clay
ooze

calcareous ooze
siliceous ooze
reefs
fringing reefs

barrier reefs
atoll
ophiolites
Exclusive Economic Zone

CHAPTER CONCEPT QUESTIONS

1. Where did sea water come from, and how old are the oceans?

2. Briefly summarize the contributions to our understanding of the oceans of the following: The Deep Sea Drilling Project, submersibles such as *Alvin,* echo sounding, seismic profiling.

3. Compare and contrast the continental rise, slope and shelf in their slopes, depths, and sedimentary processes.

4. How does a turbidity current work? What is the evidence that they exist?

5. Compare and contrast the two types of continental margins.

6. Even though the sea floor is rough when it is created at the ridge, the abyssal plains are flat. Explain why.

7. Summarize the characteristics of the oceanic trenches. Include information on typical depths, slopes, heat flow, gravity anomalies, and seismic activity.

8. Summarize the characteristics of the oceanic ridges.

9. Compare and contrast seamounts, guyots, abyssal hills, and aseismic ridges.

10. List the various types of sediment found in the deep sea and state the origin of each.

11. What is are ophiolites? What have they contributed to our knowledge of the oceanic crust.

12. What resources are currently provided by the ocean. What resources may be supplied in the future? What special problems does extraction of these resources pose?

CHAPTER SUMMARY

1. The oceans cover about _____ percent of the Earth.

2. The oceans formed at least _____ years ago by the condensation of volcanic-derived _____.

3. During the 15 years of the Deep Sea Drilling Project, the _____ drilled over 1000 holes in the sea floor.

4. The subsurface geology of the sea floor is determined by _____ profiling.

5. Continental margins consist of the continental _____, the continental _____, and continental _____.

6. The gradient of the shelf is about _____ degrees, while that of the slope averages about
    _____ degrees.

7. In some places, the shelf and slope have been eroded, forming submarine _____.

8. Density currents called _____ currents have transported sediment from the continental shelf
    and slope to the deep sea floor. This sediment has largely accumulated in overlapping _____
    at the continental _____.

9. Turbidites have also been deposited beyond the continental rise to form the _____ plains.

10. _____ continental margins occur on the leading edge of a continent where oceanic crust is
    subducted in a deep oceanic _____.

11. Active continental margins are characterized by narrow continental _____ and the lack of a
    continental _____.

12. Passive continental margins occur on the _____ edges of continental plates.

13. Vast, flat _____ commonly occur at the base of the continental rise.

14. The deepest parts of the oceans are found in the deep sea _____. The deepest of these exceed
    _____ feet deep.

15. Much of the _____ Ocean is rimmed by deep oceanic trenches.

16. The mid-oceanic ridges form a mountain range that extends _____ km around the globe.

17. Running along the crests of some oceanic ridges are _____ formed in response to tensional forces.

18. Ocean ridges are _____ by sea floor fractures.

19. Here and there, rising over a kilometer above the flat abyssal plain are _____. Those with a
    flat top are called _____.

20. Ocean ridges that lack earthquakes are called _____ ridges.

21. Deep sea sediments include large amounts of fine-grained material derived from the continents and
    islands called _____ clay.

22. _____ is the name given to sediments rich in the shells of microscopic plankton. These
    deposits can be _____ or siliceous in composition.

23. _____ are structures built by the skeletons of colonial organism.

24. In an island setting reefs evolve from _____ reefs to _____ reefs to atolls.

25. _____ are slivers of sea floor emplaced on continental crust above subduction zones.

26. The _____ Zone was designed to proclaim sovereign rights over these resources.

MULTIPLE CHOICE

1. Submersible submarine which has contributed greatly to our understanding of the oceanic ridges.
    a. H.M.S. *Beagle*
    b. *Glomar Challenger*
    c. JOIDES *Resolution*
    d. *Alvin*

2. The steepest slopes of the continental margins occur on the
    a. continental shelf.
    b. continental rise.
    c. continental slope.
    d. abyssal plain.

3. The internal character of the oceanic crust beneath the sea floor has been determined by
    a. deep drilling.
    b. seismic profiling.
    c. gravity surveys.
    d. all of the above

4. Submarine canyons cut across the
    a. abyssal plains.
    b. ocean ridges.
    c. trenches.
    d. none of the above

5. Submarine canyons were, in part formed by
    a. stream erosion when sea level was lower than it is today.
    b. stream erosion when sea level was higher than it is today.
    c. rifting at mid-ocean ridges.
    d. wind erosion.

6. Turbidity currents deposit sediments on
    a. submarine fans and the abyssal plains.
    b. the abyssal plains and ridge crests.
    c. the continental shelf and continental slope.
    d. none of these

7. Black smokers are found associated with
    a. passive continental margins.
    b. active continental margins.
    c. oceanic ridges.
    d. abyssal plains.

8. A seamount with a flat top is called a(n)
    a. guyot.
    b. atoll.
    c. abyssal hill.
    d. aseismic ridge.

9. Passive continental margins are associated with
    a. mountains.
    b. volcanoes.
    c. high seismic activity.
    d. none of the above

10. The abyssal plains are flat because
    a. they were eroded flat during low sea level stand.
    b. they are covered with flat-lying sediment.
    c. they are capped by flat basalt flows.
    d. all of the above

11. Deep sea trenches are most common in the
    a. Pacific Ocean.
    b. Atlantic Ocean.
    c. Indian Ocean.
    d. Arctic Ocean.

12. The trenches have
    a. low heat flow and low seismic activity.
    b. high heat flow and high seismic activity.
    c. low heat flow and high seismic activity.
    d. huge positive gravity anomalies.

13. Most of Earth's intermediate and deep focus earthquakes are associated with
    a. mid-ocean ridges.
    b. deep sea trenches.
    c. abyssal plains.
    d. passive continental shelves.

14. A rift valley extends
    a. along an ocean ridge crest.
    b. along the floor of a deep sea trench.
    c. across most abyssal plains.
    d. none of these

15. Sea floor fractures produce offsets in
    a. submarine canyons.
    b. trenches.
    c. guyots.
    d. ridge crests.

16. The most abundant type of sediment in the deep sea is
    a. cosmic dust.
    b. gravel from rivers.
    c. pelagic clay and ooze.
    d. deposits resulting from reaction with sea water.

17. Ooze is sediment that
    a. is carried to the continental shelves by streams during periods of low sea level.
    b. sediment that is carried by icebergs then dropped when the ice melts.
    c. clay formed from the shells of planktonic organisms.
    d. sediment that is blown out to the middle of the ocean, then settles gradually to the bottom.

18. Reefs separated from an island by a lagoon are called
    a. patch or pinnacle reefs.
    b. atolls.
    c. barrier reefs.
    d. fringing reefs.

19. Which of the following is NOT part of an ophiolite sequence?
    a. pillow lavas
    b. layered gabbro
    c. massive rhyolite
    d. deep sea sediments

20. Manganese nodules may prove to be a valuable source of _____ in the future.
    a. manganese
    b. cobalt
    c. nickel
    d. all of these

TRUE OR FALSE

___1.   The ancient Greeks discovered most of the features of the sea floor almost 2000 years ago.

___2.   The oceans are at least 3.5 billion years old.

___3.   Echo sounding is used to determine the subsurface geology of the ocean basins.

___4.   The edge of the continental crust does not necessarily correspond to the shorelines.

___5.   During the Pleistocene, sea level was generally higher than today.

___6.   The flattest part of the continental margin is on the continental shelf.

___7.   Passive continental margins usually lack a continental rise.

___8.   The west sides of South and North America represent passive continental margins.

___9.   Along the bottom of an ocean basin, sea water has a temperature very close to zero degrees Centigrade.

___10.   The deepest parts of the ocean occur in the abyssal plains.

___11.   Gravity surveys indicate that trenches are not in isostatic equilibrium.

___12.   Sea floor forms at a mid-ocean ridge.

___13.   Aseismic ridges are characterized by intensive seismic activity.

___14.   Gabbro is an important constituent of the oceanic crust.

___15.   Coral reefs require clear, warm, shallow water.

___16.   Subsidence plays a necessary role in the formation of guyots and atolls.

___17.   Fringing reefs are attached to an island or continent.

___18.   Oozes are deep sea sediments rich in windblown dust and volcanic ash derived from the continents.

___19.   Over 15% of U.S. oil production is from continental shelf areas.

___20.   The EEZ was a region set aside as a sanctuary for marine life.

DRAWINGS AND FIGURES

1.   Sketch a series of 3 diagrams depicting the transition of an island from and active volcanic island with a fringing reef to and inactive volcanic island with a barrier reef to an atoll.

2.   Draw a cross-section of the sea floor emphasizing the various layers of the oceanic crust. Use Fig. 11-24 to check your answer.

CHAPTER 12

PLATE TECTONICS: A UNIFYING THEORY

CHAPTER OBJECTIVES

By the end of this chapter you should be able to:

1. List some of the early evidence that the continents of Africa and South America were once together, which provided the seeds of what was to become the continental drift hypothesis.
2. List the evidence that Wegner used to argue for his hypothesis of continental drift.
3. Discuss how the pioneering work in paleomagnetism of S. K. Runcorn supported the hypothesis of continental drift.
4. Summarize the theory of sea floor spreading as postulated by Harry Hess and list his evidence for it.
5. Summarize the paleomagnetic evidence for sea floor spreading contributed by Vine, Matthews and Morley.
6. Summarize the evidence in favor of sea floor spreading contributed by the Deep Sea Drilling Project.
7. Characterize the different types of plate boundaries recognized in plate tectonic theory.
8. Discuss several ways in which the past and present rate and direction of plate movement can be determined.
9. Summarize the alternative mechanisms for plate movement currently being discussed by geologists.
10. Rationalize the apparent relationship between plate tectonic boundaries and valuable metallic mineral deposits.

A USEFUL ANALOGY

Our Eatable Earth

You can build your own subduction zone to illustrate the accretion of a subduction complex to a continental margin. Find two tables that are not quite the same height. The two tables must be almost the same height so that the table tops overlap in thickness. In other words, as you push a sheet of cardboard from the shorter table toward the taller table, the cardboard will not be able to slide onto the top of the higher table but will not slide under the table top either; it will slide into the side of the table top. Set these two tables beside each other so that there is barely a gap between them.

Now, take a table cloth and put it on the shorter table so that it hangs between the tables. The shorter table is oceanic crust, the taller is continental crust, and the tiny gap between them is the trench.

Spread some peanut butter across the top of the oceanic crust getting plenty in the trench. This is sediment deposited just off the edge of the continent. Send some worthy victim, such as a younger sibling, below to pull the table cloth down through the trench. This is the oceanic crust moving beneath the continental crust. As this happens, the sediment (peanut butter) will not be able to go down the trench and will pile up in a highly contorted manner at the edge of the continent....instant subduction complex! You can vary the experiment by adding pie or cakes as microcontinents to raft into the continent.

KEY TERMS

After reading this chapter, you should be familiar with the following terms.

Gondwana	transform plate boundary
Glossopteris flora	spreading ridges
continental drift	subduction
Pangaea	Benioff Zone
Laurasia	oceanic-oceanic boundary
Mesosaurus	subduction complex
Lystrosaurus	paleographic maps
Cynognathus	back-arc basin
polar wandering	oceanic-continental boundary
sea-floor spreading	continental-continental
thermal convection cells	mélange
Glomar Challenger	ophiolite
plate tectonics	transform faults
plates	hot spots
divergent plate boundary	ridge push
convergent plate boundary	slab pull

CHAPTER CONCEPT QUESTIONS

1. Summarize the evidence cited by Wegner as support for the hypothesis of continental drift. Include evidence suggested by map patterns, fossils, rock sequences, mountain ranges, and Paleozoic placation.

2. If so much evidence had accumulated in support of the original idea of continental drift, why did geologists reject this idea?

3. What are the alternative explanations for the variation in the orientation of magnetic minerals in rocks of different ages? Which of these is the best explanation and why?

4. How does the theory of sea floor spreading differ from continental drift? What new observations in the 1950s lead to the development of this theory?

5. What do the magnetic "stripes" across the sea-floor indicate? How do they support the theory of sea floor spreading?

6.   Compare the age of the oceanic crust to the age of the continental crust. How does this data support sea floor spreading?

7.   Summarize succinctly but completely, the characteristics of divergent plate boundaries.

8.   Enumerate the steps in the rifting of a continent and the opening of an ocean basin.

9.   What geologic features can be used by geologists to identify sites of ancient rifting?

10. Compare and contrast the characteristics of ocean-ocean, ocean-continent, and continent-continent convergent margins.

11. What geologic features can be used by geologists to identify sites of ancient convergent plate boundaries?

12. What are ophiolites? What have they contributed to our knowledge of the oceanic crust?

13. How can the speed and direction of motion of a plate be determined?

14. Describe the alternative driving mechanisms for plate tectonics?

COMPLETION QUESTIONS

1.   When the present continents were joined together, they formed the super continent of _____, with the northern hemisphere continents forming _____, and the southern hemisphere continents forming _____.

2.   The abundant fossils of the _____ flora were used as evidence by early proponents of continental drift.

3.   The father of continental drift is generally considered to be _____.

4.   The orientation of glacial striations indicate that the late Paleozoic south pole was located in present-day _____.

5.   The discovery of the great extent of the mid-ocean ridge systems prompted Harry Hess to propose the theory of _____.

6.   The symmetric disposition of alternating positive and negative _____ about the _____ is one of the most convincing pieces of evidence for the theories of sea floor spreading and plate tectonics.

7.   The oldest sea floor is less than _____ years old and the oldest continental crust is _____ old.

8.   The _____ Project confirmed earlier observation on the paleomagnetism of the sea-floor plus added information on the age and thickness of the sediments.

9.   Sea floor spreading states that sea floor _____ at an oceanic ridge, and is _____ in a trench.

10. Plate tectonics states that plates are large slabs of _____ which float on semiplastic _____.

11. As a result of sea floor spreading, with increasing distance form a ridge crest, the sea floor becomes _____ in age, the sediments become _____ in age, and the sediments become _____ in thickness.

12. The main rock type produced at oceanic ridges and, therefore, making up the sea floor is _____.

13. When a continent undergoes extension, a large linear depression called a _____ forms. A good example of this occurs in Eastern Africa.

14. Convergent boundaries are characterized by dipping planes of earthquake foci called _____ zones.

15. As sea-floor is subducted, it partially melts. The resulting magma rises to the surface to produce volcanic rocks which are _____ in composition.

16. A curved chain of islands called an _____ forms over subduction zones at ocean-ocean convergent margins.

17. The lithosphere on the landward side of a volcanic island arc may be stretched and thinned resulting in a _____.

18. As a plate is subducted, sediments are scraped off the descending plated at the trench and added to the overriding plate as a _____ complex.

19. Ancient convergent boundaries are sometimes recognized by the presence of highly deformed and metamorphosed submarine rocks known as _____.

20. At a transform boundary, two plates slide past one another, as shown by the _____ Fault in California.

21. Hot _____ are plumes of rising _____, coming from deep within the Earth.

22. The volcanoes on the Big Island of _____ have formed from a hot spot.

23. The driving mechanism for plate tectonics is not completely known, but _____ cell generated movement is a very popular theory.

MULTIPLE CHOICE

1. The earliest observation which would ultimately contribute to the body of evidence supporting plate tectonic theory was
   a. *Glossopteris* fossils on the southern continents.
   b. the apparent fit of South American and African coasts.
   c. the apparent continuity of mountain ranges across the Atlantic.
   d. magnetic anomalies on the sea floor.

2. All of the present day continents were at one time joined together to form
   a. Gondwanaland.
   b. Laurasia.
   c. Pangaea.
   d. none of the above

3. Glacial evidence from _____ supports the concept of Pangaea.
   a. the northern hemisphere
   b. the southern hemisphere
   c. Laurasia
   d. the Arctic

4. Glossopteris was
   a. a seed fern.
   b. an early mammal.
   c. type of dinosaur.
   d. none of the above

5. Which of the following was not part of Gondwanaland?
   a. Africa
   b. Asia
   c. South America
   d. Australia

6. _____ proposed the theory of _____ in the early 1960s and this led very quickly to the theory of plate tectonics.
   a. Alfred Wegner / continental drift
   b. Alfred Wegner / sea floor spreading
   c. Harry Hess / continental drift
   d. Harry Hess / sea floor spreading

7. The best explanation for polar wandering curves is
   a. that the continents were fixed as the north magnetic pole shifted position.
   b. that there were two north magnetic poles over much of Earth's history.
   c. that the north magnetic pole was fixed and the continents moved.
   d. none of these

8. In a convection cell _____.
   a. hot, less dense material rises upward
   b. hot, more dense material rises upward
   c. cool, less dense material sinks downward
   d. cool, more dense material rises upward

9. The magnetic anomalies about a ridge crest are due to
   a. alternating basalt and rhyolite flows.
   b. thick accumulations of magnetite rich sediment.
   c. manganese nodules.
   d. reversing polarity of the Earth's magnetic field.

10. Compared to oceanic crust, continental crust is
    a. younger.
    b. older.
    c. about the same age.
    d. cannot accurately be dated.

11. The newest oceanic crust, and the thinnest deep sea sediments are found at
    a. deep sea trenches.
    b. ridge crests.
    c. transform faults.
    d. abyssal plains.

12. Convergent plate boundaries form where
    a. two continent bearing plates collide.
    b. two ocean bearing plates collide.
    c. an oceanic and a continental bearing plate collide.
    d. any of the above

13. Transform plate boundaries
    a. occur when two plates slide horizontally past one another.
    b. are exemplified by the San Andreas Fault.
    c. can offset ridge crests.
    d. all of the above

14. Trenches are NOT present at _____ convergent margins.
    a. ocean-ocean
    b. ocean-continent
    c. continent-continent
    d. any

15. The chaotic mixture of rock known as melange is indicative of a _____ plate boundary.
    a. divergent
    b. convergent
    c. transform
    d. any of these

16. The fastest moving plate is
    a. the Pacific plate.
    b. the North American Plate.
    c. the South African Plate.
    d. the Arabian Plate.

17. The Aleutian Islands, the Tonga arc, the Japanese and the Philippine Islands are an example of
    a. an oceanic and a continental bearing plate colliding.
    b. two continental bearing plates colliding.
    c. two ocean bearing plates colliding.
    d. hot spot formed island chains.

18. The Peru-Chile Trench and the Andes Mountains are an example of
    a. an oceanic and continental bearing plate colliding.
    b. two continental bearing plates colliding.
    c. two ocean bearing plates colliding.
    d. a transform fault.

19. The Himalayas, Appalachians and Alps are an example of
    a. an oceanic and continental bearing plate colliding.
    b. two continental bearing plates colliding.
    c. two ocean bearing plates colliding.
    d. a transform fault.

20. Most of the worlds trenches border
    a. the Atlantic Ocean.
    b. South America.
    c. the Indian Ocean.
    d. none of the above

21. The Hawaiian Islands are an example
    a. an oceanic and a continental bearing plate colliding.
    b. two continental bearing plates colliding.
    c. two ocean bearing plates colliding.
    d. a hot spot formed island chain.

TRUE OR FALSE

___1.   Because geologists are not completely clear on the driving mechanism for plate tectonics, this theory has as many opponents as proponents.

___2.   Only the northern hemisphere continents formed the continent of Pangaea.

___3.   When reconstructing Pangaea, the best reconstruction is given by matching present day shorelines.

___4.   Glossopteris flora are found in all of the southern hemisphere continents.

___5.   Sea floor spreading is a process in which sea floor forms at a ridge crest and is destroyed in a trench.

___6.   In plate tectonics, the continents move laterally by pushing their way through the oceanic crust.

___7.   Batholiths commonly occur at ocean-continent convergent margins.

___8.   The highest mountains in the world are at ocean-continent convergent margins.

___9.   The symmetrical pattern of magnetic anomalies about the axes of trenches, finally proved sea floor spreading.

___10. With increasing distance from a ridge crest, the sea floor becomes older and the sediments become thicker.

___11. Hawaii is an example of an island arc.

___12. Most geologists now agree that the main mechanism of plate movement is the "ridge-push".

___13. The East African Rift Valley illustrates a continent in the early stages of fragmentation.

___14. When an ocean bearing plate collides, with a continent bearing plate, the ocean bearing plate always subducts beneath the continent bearing plate.

___15. The Sea of Japan is a good example of a back arc basin.

___16. Plate velocity can be determined using magnetic anomalies.

___17.   Most hot spots occur at plate margins.

___18.   Transform boundaries tend to have little or no earthquake activity.

___19.   As of yet, there has been no demonstrable relationship between mineral resources and plate tectonics.

___20.   The Atlantic Ocean is currently shrinking.

DRAWINGS AND FIGURES

1.   For each plate boundary shown on the map below, indicate by arrows which are convergent and which are divergent. Draw arrows pointing toward each other at convergent boundaries, and away from each other at divergent boundaries.

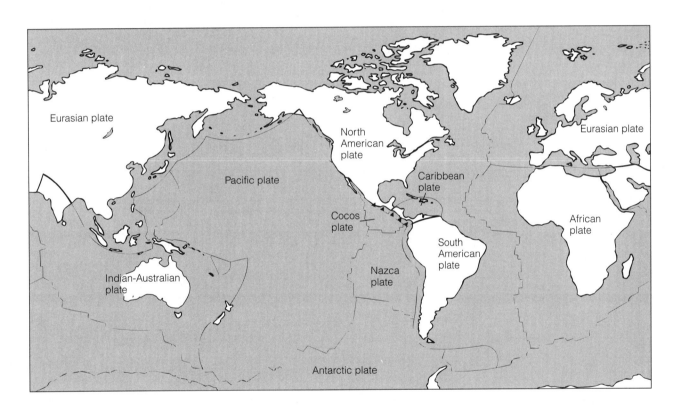

2.   What type(s) of plate margin(s) is(are) shown in the block diagrams below? Label (a) a rift valley, (b) the continental shelf(s), (c) fault blocks, and (d) the mid-ocean ridge.

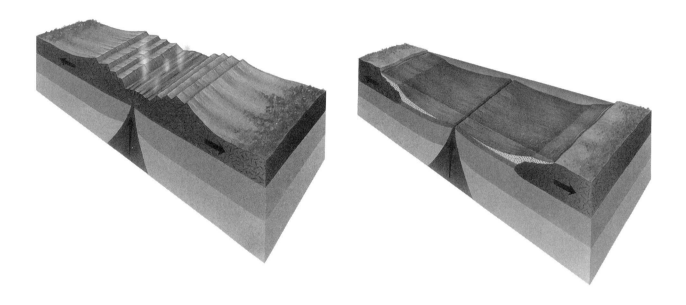

3.   What types of plate margins are shown in the block diagrams below? Label the (a) volcanic
     arc(s), (b) the trench(es), (c) the back-arc basin(s), and (d)the subduction zone(s)

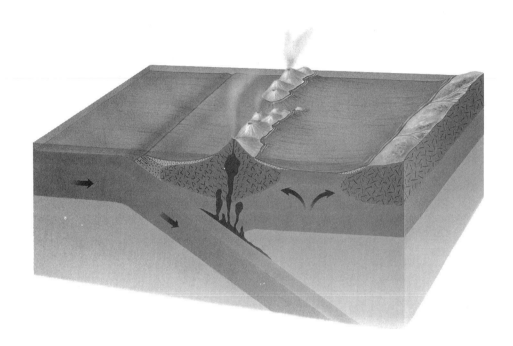

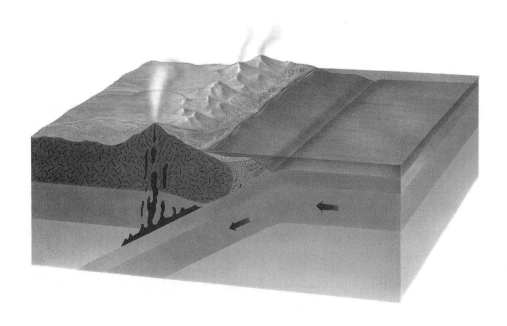

CHAPTER 13

DEFORMATION, MOUNTAIN BUILDING, AND
THE EVOLUTION OF CONTINENTS

CHAPTER OBJECTIVES

By the end of this chapter you should be able to:

1. Differentiate between stress and strain and define the various categories of stress and strain.
2. Describe the orientation of a tilted plane such as a bedding plane or a fault plane.
3. Name the parts of a fold and classify folds.
4. Differentiate between joints and faults and classify faults.
5. Describe several ways mountains are formed.
6. Define orogenesis and describe several processes that take place during orogenesis.
7. Compare orogenesis along the various types of convergent margins and give examples of each.
8. Explain how continents grow by accretion and state the evidence for that process.
9. Differentiate between shields, platforms, and cratons.
10. Be familiar with the major events in the evolution of North America from Precambrian times to the present day.

USEFUL ANALOGIES

It is easy to look at a cross-section of a fault such as those in question 3 above and figure out which block is the hanging wall and which is the footwall.

Draw a little vertically standing stick figure man across the fault so that the fault cuts him at the waste. If you draw it properly, the man's head will be in one block and his feet in the other. The block hanging over his head is the hanging wall and that in which his feet sit is the footwall.

There are many materials which you can use to demonstrate how various factors influence the type of strain that a material will show. One of the best is Silly Putty®.
When you pull the putty slowly it behaves in a plastic manner but when you snap it quickly it breaks. Therefore, the rate at which stress is applied is important in determining how Silly Putty® deforms.

A sheet of paper provides another example. Compress it and it folds (behaves plastically) but pull it and it tears (behaves brittlely). Therefore, the direction of applied stress is important in determining how paper deforms. However, it doesn't matter the rate at which you apply the stress to paper because it will behave the same way regardless.

KEY TERMS

After reading this chapter, you should be familiar with the following terms:

deformation	fault plane
stress	fault scarp
strain	fault breccia
compression	hanging wall block
tension	footwall block
shear stress	dip-slip fault
elastic strain	normal fault
plastic strain	reverse fault
fracture	thrust fault
brittle	strike slip fault
ductile	left-lateral
strike	right-lateral
dip	geologic map
folds	oblique-slip fault
monoclines	mountain
anticline	mountain range
syncline	mountain systems
axial plane	block faulting
limb	horst
symmetrical fold	graben
asymmetrical fold	orogeny
overturned fold	orogensis
recumbent fold	orogen
fold axis	continental accretion
plunging fold	accretionary wedge
dome	microplate
basin	shields
joints	craton
fault	

CHAPTER CONCEPT QUESTIONS

1. What is the difference between stress and strain?

2. What is the difference between brittle and plastic strain? What factors determine if rocks will be plastic or brittle in their behavior? What geologic structures result from brittle vs. ductile behavior?

3. Explain how the orientation of a plane is described using strike and dip.

4. Define, in terms of fold limbs, axes, and axial planes, the following types of folds: (a) symmetrical, (b) asymmetrical, (c) overturned, (d) recumbent, (e) nonplunging, (f) plunging.

5. Compare and contrast an anticline and a dome (or between a syncline and a basin)?

6. Compare and contrast a joint and a fault?

7. Define, in terms of hanging wall and footwall, (a) normal fault, (b) reverse fault, and (c) strike-slip fault.

8. Under what condition is a reverse fault considered a thrust fault?

9. What kind of forces cause normal and reverse faulting?

10. How can you tell if a fault is a left-lateral or a right-lateral strike slip fault?

11. Briefly describe four different ways that mountains form.

12. Compare and contrast orogensis along an ocean-ocean convergent boundary with an ocean-continent convergent boundary and a continent-continent convergent boundary. Give an example of each.

13. What geologic evidence suggests that parts of some continents (the west coast of North America, for example) were transported in from distant parts of Earth?

14. Compare and contrast an shield, a platform and an orogen.

15. Why are highly deformed metamorphic rocks such a common part of the continental crust, even under shields and platforms far from the plate margins?

16. Construct a time line which highlights the major geologic events in the evolution of North America. Include the following: A) 2.0 to 1.8 billion years ago; B) Late Paleozoic (east); C) Late Triassic (east); D) Cretaceous to Eocene (west); E) Cenozoic (east and west). You may need to refer to Figure 8.2 to refresh your memory about time periods and their translation to number of years before present.

COMPLETION QUESTIONS

1. Stress is _____ per unit area.

2. Squeezing rocks creates a _____ stress, pulling rocks in opposite directions creates _____ stress, and _____ stress develops when parallel forces act in opposite directions.

3. Rocks may behave as _____ materials to produce fractures, or they may deform by plastic strain to create _____.

4. The trend of a bedding plane is called its _____, while its angle of inclination from the horizontal is called the _____.

5. An upward flexure of the rock layers forms an _____, while a downward flexure creates a _____.

6. The _____ divides a fold into two equal halves. Each half is called a _____.

7. If a fold axial plane is not vertical, the fold is said to be _____, if it is horizontal the fold is _____.

8. In a nonplunging fold, the fold axis is _____, while in a plunging fold the fold axis is _____ from the horizontal.

9. In a dome the beds dip _____ from the center in all directions, while in a basin, the beds dip _____ the center from all directions.

10. If relative movement occurs between rocks on either side of a fracture, the fracture is said to be a _____. If no movement as occurred, the fracture is a _____.

11. In a normal fault the hanging wall moves _____ relative to the footwall.

12. In a reverse fault the _____ wall moves up relative to the _____.

13. In a thrust fault the relative motion is the same as in a _____ fault, but the angle of dip of the fault plane is less than _____ degrees.

14. _____ forces are responsible for strike-slip faults.

15. Oblique faults have both _____ and _____ displacement.

16. A mountainous region consisting of many mountain ranges is called a mountain _____.

17. In areas which are being stretched due to tensional stresses, mountains can be formed by _____ faulting.

18. In areas which are being stretched due to tensional stresses, the uplifted blocks, called _____, form the mountains; and the down-dropped blocks, called _____, form valleys.

19. The deformation of the Earth's crust to form mountains is called an _____.

20. Sediments and even sea floor may be scraped off a descending plate during subduction and form a highly deformed terrain called an _____.

21. The Andes Mountains are rising where _____ bearing plate is converging with _____ bearing plate.

22. Much of the west coast of North America developed from accretion of _____, which differ from one another in terms of fossil content, paleomagnetism and structural trend.

23. Regions of exposed, ancient granitic and metamorphic crustal rocks are called _____.

24. The _____ Mountains formed during the Paleozoic in response to compression during the amalgamation of Pangaea.

25. Much of the development of the Cordillera was in response to the subduction of the _____ plate beneath the North American plate.

MULTIPLE CHOICE

1. If a rock permanently bends but does not break it is demonstrating
    a. elastic behavior.
    b. plastic behavior.
    c. brittle behavior.
    d. none of the above.

2. Plastic versus brittle rock behavior is determined by
    a. the temperature.
    b. the rock type.
    c. the type of stress.
    d. all of the above.

3. Strike is defined as
    a. the angle from a horizontal plane down to a bedding plane.
    b. the intersection of a horizontal plane with a tilted plane.
    c. the angle of the dip of the fold axis.
    d. none of these

4. The number adjacent to a strike and dip symbol on a geologic map indicates
    a. the strike of a bed.
    b. the dip of a bed.
    c. neither a nor b
    d. both a and b

5. In an anticline
    a. the beds arch upwards.
    b. the limbs dip toward each other.
    c. the youngest rocks are in the center of the fold.
    d. all of the above

6. A fold is divided into two equal halves by the
    a. fold axis.
    b. axial plane.
    c. limb.
    d. plunge.

7. If the axial plane is horizontal, the fold is classified as a
    a. recumbent fold.
    b. symmetrical fold.
    c. asymmetrical fold.
    d. plunging fold.

8. In a plunging fold the _____ is inclined from the horizontal
    a. strike
    b. axial plane
    c. fold axis
    d. none of the above

9. In a dome the beds
    a. are nearly horizontal.
    b. dip toward the center in all directions.
    c. dip vertically.
    d. dip away from the center in all directions.

10. Joints are
    a. a fracture with vertical displacement.
    b. a fracture with horizontal offset.
    c. a fracture with both vertical and horizontal offset.
    d. none of the above

11. Faults are fractures that can show
    a. vertical offset.
    b. horizontal offset.
    c. ether a or b
    d. no displacement at all.

12. The "hanging wall"
    a. overlies the fault plane.
    b. overlies the axial plane.
    c. lies beneath the fault plane.
    d. lies beneath the axial plane.

13. In a reverse fault, the hanging wall
    a. moves down relative to the footwall.
    b. moves up relative to the footwall.
    c. moves horizontally away from the footwall.
    d. remains stationary.

14. If you are standing by a right-lateral strike-slip fault, looking across the fault plane the displacement of the opposite block appears to be
    a. up.
    b. down.
    c. horizontally to the left.
    d. horizontally to the right.

15. The San Andreas in California is a
    a. large right-lateral strike-slip fault.
    b. large left-lateral strike-slip fault.
    c. large normal fault.
    d. large thrust fault.

16. Fault block mountains are due to
    a. thrust faulting.
    b. reverse faulting.
    c. strike slip faulting.
    d. normal faulting.

17. Horsts form _____ while grabens form _____.
    a. valleys / mountains
    b. normal faults / reverse faults
    c. mountains / valleys
    d. none of the above

18. Thrust faults result from _____ stress.
    a. compressional
    b. extensional
    c. shear
    d. none of the above

19. Which would you LEAST expect to find at an ocean-continent convergent margin?
    a. volcanic mountains.
    b. thrust faulted mountains.
    c. accretionary wedge.
    d. You find ALL of them.

20. A continental-continental plate collision forms
    a. a trench.
    b. a mountain range.
    c. a chain of volcanic islands.
    d. all of the above

21. The Rocky Mountains are part of the region called the
    a. Cordillera.
    b. Appalachians.
    c. Shield.
    d. Craton.

TRUE OR FALSE

___1. The same material may behave in a ductile manner when under compression and a brittle manner under tension.

___2. Strain is deformation due to stress.

___3. Dip is the angle of inclination between an imaginary horizontal plane and a bedding plane.

___4. As rocks become more deeply buried they become more brittle.

___5. A syncline is a fold where the two limbs dip toward one another.

___6. Overturned folds always have an upside down limb.

___7. In a basin the beds dip toward the center in all directions.

___8. Unlike bedding planes, fault planes lack a strike and dip.

___9. Joints are more abundant than faults.

___10. In a normal fault, the hanging wall moves up relative to the footwall.

___11. Thrust faults are particularly shallow dipping reverse faults.

___12. Oblique-slip faults have both dip-slip and strike-slip offset.

___13. A mountain system comprises several mountain ranges.

___14. The Basin and Range Province formed by thrust faulting.

___15. Normal faults are due to tensional stresses in the crust.

___16. Most of Earth's largest mountain ranges resulted from tensional stresses.

___17. Mountain ranges which originate from compression and which are located in the middle of continents probably resulted from continent to continent collisions.

___18. Much of the west coast of North America is composed of microcontinents which originated from different parts of the earth.

___19.   Accretionary wedge sediments are associated with trenches.

___20.   Shields are areas of craton that are currently tectonically unstable.

___21.   The mountains of western North America are younger than those of eastern North America.

DRAWINGS AND FIGURES

1.   In the adjacent block sketch and label
the axial planes then label all fold
axes and limbs. Also label an anticline
and which is a syncline.

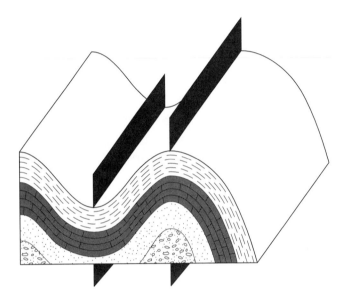

2.   Shown below are five cross-sections of folds. Identify completely the type of each. Layers are
numbered from oldest(1) to youngest(3).

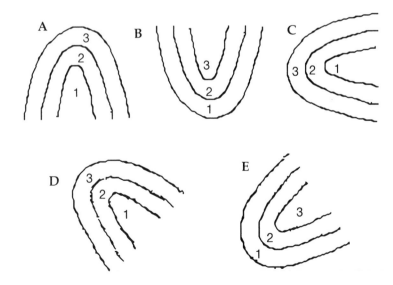

3.   Shown below are two cross-sections of faults. Label the normal fault and the reverse fault.

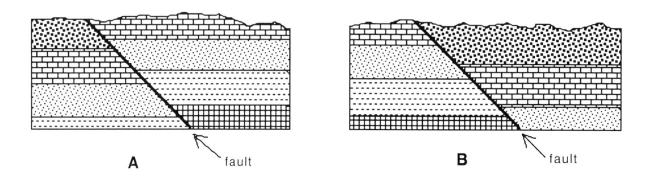

4.   The figure below is a map view of a strike-slip fault that has offset some railroad tracks. Is it right or left-lateral?

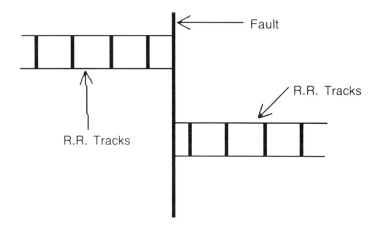

CHAPTER 14

MASS WASTING

CHAPTER OBJECTIVES

By the end of this chapter you should be able to:

1. Define mass wasting.
2. List the factors that define a slope's shear strength.
3. List and discuss the factors that influence an area's susceptibility to mass wasting.
4. List and differentiate between the different types of mass wasting.
5. Describe what sorts of activities a hazard assessment study entails.
6. Describe a number of ways the detrimental effects of mass wasting can be mitigated.

KEY TERMS

After reading this chapter, you should be familiar with the following terms:

landslide
mass wasting
shear strength
angle of repose
rapid mass movement
slow mass movement
rockfalls
talus
slide
slump
rock slide
mudflow

debris flow
earthflow
quick clays
solifluction
permafrost
creep
complex movement
debris avalanche
slope stability map
cut-and-fill
benching
rock bolts

CHAPTER CONCEPT QUESTIONS

1. What is mass wasting and why it occur?

2. What factors define a slope's shear strength?

3. What is the significance of the angle of repose?

4.   Describe how each of the following factors influences mass wasting: (a) slope angle, (b) weathering and climate, (c) water content, (d) overloading, and (e) bedrock geology of the area. For each of these, describe a situation in which that particular factor may be manipulated by man or nature so as to induce a mass movement.

5.   Distinguish between rapid and slow movement. List examples of each.

6.   Distinguish between the general categories of mass movement (i.e. falls, slides and flows).

7.   Define rockfall and list three factors which can cause it.

8.   Compare and contrast slumps and rock slides. What factors can lead to each?

9.   Distinguish between a debris flow, a mudflow, and an earthflow.

10.   What is solifluction and what factors lead to it?

11.   Why is creep so often overlooked as a process in an area? What signs might tip you off that it is occurring?

12.   When performing a hazard assessment, what signs of mass movement history and potential do geologists look for?

13.   Describe three ways in which a hillside that has a high potential for mass movement can be stabilized.

COMPLETION QUESTIONS

1.   A slope will be stable as long as its _____ strength is greater than the downslope pull of _____.

2.   The steepest angle that most slopes can maintain without collapsing (called the angle of _____) is between _____ and _____ degrees.

3.   A steep slope is generally _____ stable than a gentle slope.

4.   Water acts to reduce the _____ between the soil particles.

5.   A slope that is _____ to the dip of the underlying sedimentary bedrock layers may be particularly unstable.

6.   The most common triggering mechanism for slope failure is the shaking due to _____ and excessive amounts of _____.

7.   If the movement of material is fast enough to be visible, then it is considered to be _____ mass movement.

8.   Talus piles accumulate at the base of cliffs by the process of _____.

9.   In a _____ there is a downward movement and rotation of material along a curved surface of rupture.

10. Where a hill slope is parallel to the dip of the underlying sedimentary bedrock, _____ may occur.

11. A mass movement in which the material behaves in a viscous or plastic manner is termed a _____.

12. In a _____ a mass of material rich in silt, clay and water moves rapidly down a slope.

13. _____ flows are similar to mudflows, but are composed of larger-sized particles, contain less water and flow more slowly.

14. Earthflows move more _____ than mudflows or debris flows.

15. _____ clays spontaneously liquefy and flow when disturbed by shaking.

16. In the high latitudes and altitudes, the seasonal melting of the upper _____ can saturate soil and cause a mass movement known as _____.

17. The slowest and most common type of downslope movement is soil _____.

18. Movement of earth material that involves more than one type of mass movement is called _____ movement.

19. Information derived from a geologic hazard assessment of an area is often compiled on _____ maps as an aid to planners and engineers.

20. Many roadcuts along interstate highways are stabilized by cutting a series of step-like terraces. This process is called _____.

MULTIPLE CHOICE

1. Which of the following would be considered a type of landslide?
   a. slump
   b. rock slide
   c. debris avalanche
   d. all of the above

2. The most important stress opposing a slope's shear strength is imparted by
   a. running water.
   b. earthquakes.
   c. frost wedging.
   d. gravity.

3. A thick weathering zone
   a. does not effect a slopes stability.
   b. causes a slope to be more stable.
   c. causes a slope to be less stable.
   d. can not occur in regions of steep slopes.

4. Water can encourage mass flow by
   a. reducing friction between grains in a regolith.
   b. undercutting a steep slope.
   c. weathering bedrock to clay.
   d. all of the above

5. Removal of vegetation tends to make a slope
    a. more susceptible to mass flow.
    b. less susceptible to mass flow.
    c. steeper.
    d. increase its shear strength.

6. The most stable geologic setting occurs when
    a. the beds dip into a hillside.
    b. the beds dip in the same direction as the slope of the ground.
    c. a dense fracture pattern is present.
    d. All of the above are equally stable.

7. Overloading usually results from
    a. plate tectonics.
    b. undercutting of waves and streams.
    c. human activity.
    d. freezing and thawing.

8. Which of the following is considered to be a slow mass movement
    a. mudflow.
    b. debris flow.
    c. soil creep.
    d. rock fall.

9. Talus cones result from
    a. rock fall.
    b. rock slide.
    c. mud flows.
    d. any of the above.

10. In a slump, the earth materials
    a. move forward by sliding down bedding or fracture plane.
    b. fall from a cliff in clumps.
    c. move downward and outward along a curving rupture plane.
    d. none of the above.

11. In a rock slide materials
    a. move forward by sliding down bedding or fracture plane.
    b. fall from a cliff in clumps.
    c. move downward and outward along a curving rupture plane.
    d. none of the above.

12. Mudflows
    a. are common in arid and semiarid areas.
    b. move very rapidly.
    c. comprise mostly silt and clay sized material.
    d. all of the above.

13. One would expect solifluction to be a particularly important process
    a. where waves are undercutting sea cliffs.
    b. in alpine meadows of the Canadian Rocky Mountains.
    c. in areas of very sparse rainfall such as Death Valley, California.
    d. in a Brazilian rain forest.

14. Which of the following moves the slowest?
    a. mudflow
    b. debris flow
    c. earthflow
    d. They all move about the same speed.

15. Generally speaking, the most common form of mass movement is
    a. debris flow.
    b. slumping.
    c. mudflow.
    d. none of the above.

16. Which of the following is NOT a sign of creep.
    a. curved tree trunks
    b. a steep scarp
    c. bulging retaining walls
    d. tilted telephone poles

17. The most common type of complex movement is a combination of
    a. fall and slide.
    b. fall and flow.
    c. slide and flow.
    d. none of the above.

18. A debris avalanche is a complex movement combining
    a. fall and slide.
    b. fall and flow.
    c. either a or b.
    d. neither a nor b.

19. Which of the following types of mass movement is the most difficult to recognize in performing hazard assessments?
    a. rock slide
    b. soil creep
    c. slumping
    d. mudflow

20. One of the most effective ways to stabilize a slope is to
    a. remove all excess vegetation.
    b. increase the rate of compaction by adding water to the slope.
    c. control and remove the water going into the subsurface.
    d. none of these

TRUE OR FALSE

___1.   Mass wasting can only occur in steep terrain.

___2.   Dry loose material cannot accumulate at a steeper angle than the angle of repose.

___3.   Weathering of bedrock helps to stabilize the ground surface making it less susceptible to mass movement.

___4.   Water may trigger slope failures by reducing the friction between soil particles.

___5.   Vegetation tends to stabilize a slope by binding soil particles together.

___6.   Loading a slope with large amounts of excavated material tends to compact the subsurface and increase a slopes stability.

___7.   Earthquakes have triggered some of the Earth's most devastating mass wasting events.

___8.   Rapid mass wasting means that the earth materials move an appreciable distance over the time span of a human life.

___9.   In a slide material acts as a viscous fluid or a plastic.

___10.   Mudflows are common in arid and semiarid regions where they are triggered by heavy rainstorms.

___11.   Under cutting by rivers can trigger rock slides.

___12.   Solifluction can occur in all climates, but it is most common in the tropics.

___13.   Permafrost makes a good firm base on which a structures can be built.

___14.   Soil creep is one of the slowest downslope movements.

___15.   Curved tree trunks can be a sign of creep.

___16.   A scarp occurs at the top of a slump.

___17.   Once a landslide has occurred in an area, it generally stabilizes and is a good building site.

___18.   The occurrence of subsurface clay layers tends to make a steep slope more stable.

___19.   A good drainage system on a hillside is an effective, yet inexpensive, way to lessen a slopes susceptibility to mass wasting.

___20.   Rock bolts may prove to be effective at preventing rock slides.

DRAWINGS AND FIGURES

1.   Identify the type of mass flow illustrated in the following block diagrams.

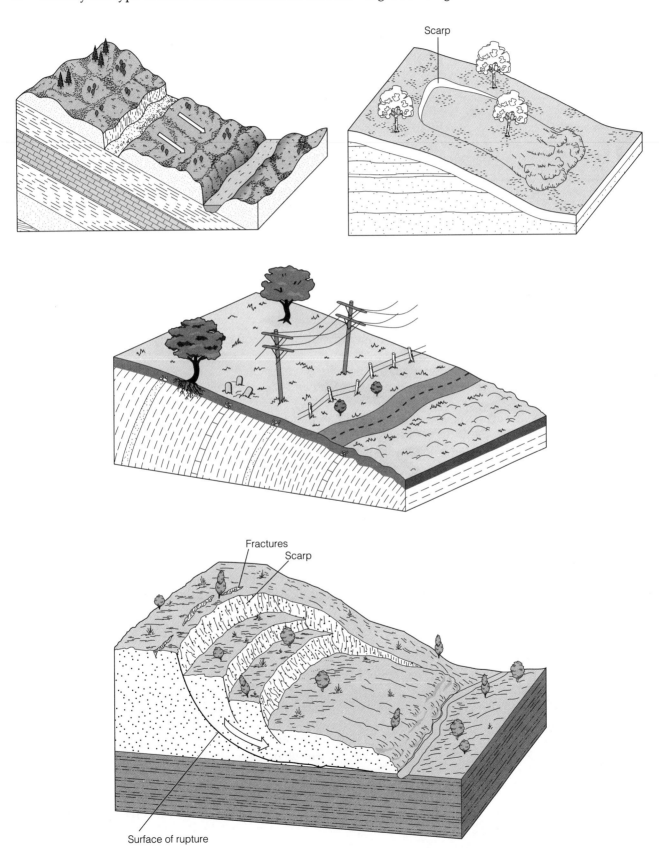

CHAPTER 15

RUNNING WATER

CHAPTER OBJECTIVES

By the end of this chapter you should be able to:

1. Say how Earth's water is distributed.
2. Reproduce the hydrologic cycle and discuss its parts.
3. Differentiate between laminar and turbulent flow.
4. Define infiltration capacity and discuss how it determines whether precipitation will result in surface water or groundwater.
5. Differentiate between sheet and channel flow.
6. Describe how velocity and discharge are determined and discuss what factors influence their values.
7. Describe how a stream erodes and what processes occur during erosion.
8. Differentiate between the different types sediment of loads that streams transport.
9. Compare and contrast braided streams and their deposits with meandering streams and their deposits.
10. Describe how delta deposition takes place and how various factors such as waves and tides can affect a delta.
11. Describe the various types of drainage patterns and for each state what factors lead to its development.
12. Explain how the tendency of a stream to seek base level and become graded affects its erosion and deposition.
13. Describe the various processes that are important in valley development.
14. Discuss how superposed streams, stream terraces, and incised meanders form.

KEY TERMS

After reading the chapter, you should be familiar with the following terms:

hydrologic cycle   transpiration
evaporation   run off
condensation   laminar flow
precipitation   turbulent flow

infiltration capacity
sheet flow
sheet erosion
channel flow stream
gradient
stream velocity
channel roughness
discharge
potential energy
kinetic energy
hydraulic action
abrasion
potholes
dissolved load
suspended load
bed load
saltation
competence
capacity
alluvium
braided streams
meandering streams
cut bank
point bar
ox-bow lake
meander scar
flood plain
lateral accretion
vertical accretion
natural leaves
recurrence interval
topset beds
delta
prograde

foreset beds
bottomset beds
topset beds
distributary channels
stream-dominated delta
bird's foot delta
wave-dominated delta
tide-dominated delta
alluvial fan
tributary streams
drainage basin
divide
drainage patterns,
dendritic drainage
rectangular drainage
trellis drainage
radial drainage
deranged drainage
ultimate base level
temporary base level
longitudinal profile
graded stream
gullies
canyon
gorge
downcutting
lateral erosion
headward erosion
stream piracy
superposed stream
water gap
stream terrace
incised meander

CHAPTER CONCEPT QUESTIONS

1. Compare and contrast evaporation, condensation, transpiration, and precipitation?

2. How does laminar flow differ from turbulent flow?

3. Explain infiltration capacity and explain how it determines whether water falling to the Earth will become groundwater or runoff.

4. How is gradient calculated?

5. Explain how gradient, channel shape and channel roughness affects stream velocity.

6. How do the concepts of potential and kinetic energy relate to a stream?

7. List and describe the different mechanisms by which streams carry sediment of different sizes?

8. What the difference between capacity and competency?

9. What factors would cause a stream to become a braided stream?

10. How do meandering streams migrate? How do oxbow lakes form?

11. How does velocity vary in a meandering river? How does it influence where erosion and deposition take place.

12. What happens when a stream overflows its banks?

13. How does a delta form? How do tides and waves modify the shape of a delta?

14. How do alluvial fans form? Why are they most abundant in arid areas?

15. What is meant by a river's drainage basin? Give five examples of ways geology influence a drainage pattern?

16. What is base level? What happens to a stream if base level is lowered?

17. How does an ungraded stream modify itself to become at least partly graded?

18. Describe three types of erosion that contribute to valley development?

19. What is stream piracy and how does it occur?

20. Detail the stages leading to the creation of a "mature" stream (such as the Mississippi River) according to the classic model. What are some shortcomings of this model?

21. Explain how each of the following forms: (a) stream terraces, (b) incised meanders, and (c) superposed streams.

COMPLETION QUESTIONS

1. Approximately _____ % of the water on Earth is in the ocean and another _____ % is tied up in glaciers.

2. About _____ % of the precipitation falling to Earth falls on land.

3. _____ flow is characterized by a mixing of stream lines.

4. The amount of rain that will become runoff is determined by the ground's _____, the maximum rate at which water can be absorbed.

5. When runoff begins, water flows in a thin film called _____ wash. Sooner or later it becomes confined in long trough-like depression called a _____.

6. The gradient of a stream is calculated by dividing its _____ by the horizontal distance that it has traveled.

7. The velocity of a stream is determined dividing the horizontal distance a stream by a unit of _____.

8. For maximum velocity, the most efficient cross-sectional shape of a channel is a _____.

9. Channel roughness influences flow velocity because of the _____ between the particles on the bottom of the channel and the flowing water.

10. _____ is the total volume of water in a stream moving past a particular point in a given period of time.

11. Only ____ % of the total _____ energy of a stream is available for erosion and transport because most of it is taken up by the internal friction of the flowing water.

12. Sedimentary particles are set in motion by the _____ action of the flowing water.

13. Potholes in solid bedrock in the beds of streams are one manifestation of the process of _____.

14. Streams carry sand and gravel as _____ load, silt and clay particles as _____ load, and ions in solution as _____ load.

15. _____ refers to the maximum-sized particles a stream can carry, and _____ refers to the total amount of sediment a stream can carry.

16. Braided streams have diverging and rejoining _____, and form if the stream is carrying excessive _____.

17. Meanders can be abandoned, forming _____ lakes or _____ scars.

18. Erosion occurs on the outside curve of a meander called the _____ and deposition on the inside of the channel where a _____ is formed.

19. When streams overflow their banks, fine-grained sediments are deposited on the _____.

20. During flooding, a ridge of sediment called a _____ is deposited along the edge of a channel.

21. Deltas form where a sediment-laden stream enters a _____ body of water, its velocity _____ and its load is deposited.

22. In a stream dominated delta, the delta _____ far seaward.

23. The diverging streams in a delta are called _____.

24. The Mississippi River Delta is a _____ dominated delta with a _____ foot shape.

25. An _____ forms where a sediment laden stream exits the mouth of a canyon and its load is quickly deposited on the valley floor.

26. The entire area from which a stream receives its discharge is called the _____ basin. Drainage _____ form the boundaries of this area.

27. A common drainage pattern resembling the veins in the leaf of a tree is called _____.

28. A _____ pattern is characteristic of a stream system flowing off a volcanic peak.

29. A rectangular drainage pattern may develop if the bedrock in the area has a pattern of right angle intersecting _____.

30. The lowest level to which a stream can erode its channel is _____.

31. The ultimate base level is _____.

32. _____ streams have a delicate balance between gradient, discharge, flow velocity, and channel characteristics such that neither significant erosion or deposition occurs in the stream channel.

33. Valleys are made deeper by stream _____ and made longer by _____.

34. Stream terraces form by a lowering of _____ which causes renewed downward erosion.

35. Deep meandering channels cut into bedrock are called _____ .

36. _____ streams cut right through resistant geologic structures.

MULTIPLE CHOICE

1. Water in a natural streams
   a. move in laminar flow.
   b. move in turbulent flow.
   c. both a and b
   d. neither a nor b

2. Which of the following has the highest infiltration capacity?
   a. A clay soil which has been compacted by herds of cattle
   b. water saturated clay soil
   c. dry, loose gravel
   d. a tar coated parking lot

3. Compute the average gradient of a stream of 3 km length which has a source at an elevation of 3000 meters above sea level and is 600 meters above sea level at its mouth.
   a. The gradient can not be computed due to insufficient information.
   b. 3000 m/1000 m
   c. 800m/km
   d. 800 km/m

4. The velocity of a stream is controlled by the
   a. discharge.
   b. channel shape.
   c. channel roughness.
   d. all of the above

5. A stream flows fastest in a channel which is
   a. deep and narrow.
   b. shallow and wide.
   c. semicircular in cross-section.
   d. Channel shape is not a factor in stream velocity.

6. A stream has a cross-sectional area of 100 square meters and is flowing at 2 meters/second. what is its discharge?
   a. 200 cubic meters/second
   b. 50 cubic meters/second
   c. 200 square meters/second
   d. 50 square meters/second

7. Which of the following types of stream load spends the most time in contact with the bottom?
   a. dissolved load
   b. suspended load
   c. bed load
   d. They all spend the same amount of time.

8. Capacity is
   a. the total amount of water a stream is carrying.
   b. the maximum-sized particle stream can carry.
   c. the total amount of load a stream can carry.
   d. none of the above

9. Braided streams are characterized by
   a. excessive bedload.
   b. point bars.
   c. lateral accretion.
   d. all of the above

10. Meanders can
    a. become deeply incised in narrow canyons.
    b. migrate back and forth across a valley.
    c. be cut off to form ox-bow lakes.
    d. all of the above

11. A natural levee is located between
    a. the channel and floodplain.
    b. a delta and the oceans.
    c. a point bar and the channel.
    d. an alluvial fan and mountain front.

12. Deltas which prograde the farthest into a lake or sea are usually
    a. stream-dominated.
    b. tide-dominated.
    c. wave-dominated.
    d. alluvial fan dominated.

13. Stream channels traversing the top of a delta are called
    a. tributaries.
    b. distributaries.
    c. secondary channels.
    d. braided channels.

14. The steeply dipping front of a delta is formed from
    a. distributaries.
    b. topset beds.
    c. bottomset beds.
    d. foreset beds.

15. The area from which a stream and its tributaries receive their discharge is called the
    a. drainage basin.
    b. drainage divide.
    c. flood plain.
    d. alluvial fan.

16. Dendritic drainage patterns develop where
    a. many intersecting fractures occur.
    b. high, isolated mountain peaks exists.
    c. where alternating resistant and nonresistant tilted beds occur.
    d. none of the above

17. Stream valleys are lengthened by
    a. headward erosion.
    b. lateral accretion.
    c. downcutting.
    d. sheet runoff.

18. Stream terraces form
    a. by downward erosion due to a rise in base level.
    b. by downward erosion due to a lowering of base level.
    c. where the velocity decreases and the stream load is deposited.
    d. none of the above

19. The ultimate base level of for the earth in general is
    a. sea level.
    b. approximately 150 meters above sea level.
    c. the bottom of the deepest abyssal plains.
    d. none of the above

20. In general, incised meanders form
    a. by an uplift of the land, and downward erosion.
    b. by headward erosion.
    c. where a stream's velocity decreases and its load is deposited.
    d. where extensive lateral erosion can occur.

TRUE OR FALSE

___1.   In laminar flow, water particles move in parallel paths.

___2.   Runoff occurs when infiltration capacity is exceeded.

___3.   Discharge is defined as the total amount of load carried by a stream.

___4.   Although oceans make up seventy percent of the Earth's surface, over half the precipitation falls on land.

___5.   The steep mountain tributaries to large rivers generally flow faster than the large rivers themselves.

___6.   Water stored behind a dam is an example of kinetic energy.

___7.   Silts and clays are transported as bed load.

___8.   Sediment transported by saltation spends much of the time on the bottom.

___9.   Braided streams are common on alluvial fans and in streams fed by glacial meltwater.

___10.   As a stream rounds a bend, there is erosion on the inside and deposition on the outside of a bend.

___11.   An ox-bow lake develops when a meander is cut off.

___12.   Floodplain sediments are typically dominated by coarse sand and gravel.

___13.   A delta's topset beds are deposited by distributary channels.

___14.   The Mississippi River Delta is a stream-dominated delta.

___15.   Alluvial fans develop best in arid and semiarid regions.

___16.   Mudflows are common features on alluvial fans.

___17.   Base level is the lowest level to which a stream can erode its channel.

___18.   Radial drainage forms in a region with numerous fractures.

___19.   In the evolution of a stream valley according to the classic model, the deepening of the valley by downcutting is followed later by the lateral erosion and valley widening

___20.   Incised meanders form by stream piracy.

___21.   Superposed streams can cut right across resistant bedrock ridges.

DRAWING AND FIGURES

1.   Draw a diagram of the hydrologic cycle. Include in your diagram condensation, precipitation to ocean and land, evaporation from ocean and land, transpiration, runoff, and groundwater. Fig. 15-5 may give you some ideas, although your version does not need to be so fancy.

2. Using the adjacent graph, answer the following questions.
    a. How fast must a current be flowing to erode sand 1/16 mm in diameter? clay 1/256 mm in diameter? gravel 2 mm in diameter.
    b. As a stream slows down, at what velocity will each of the above sediments be deposited?

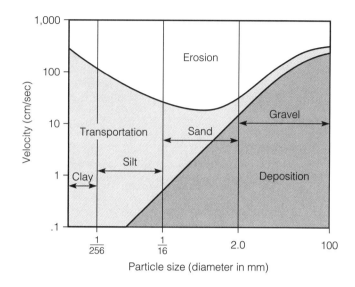

3. In the adjacent diagram, show by an arrow the path of the fastest flow. Label where erosion and deposition are taking place.

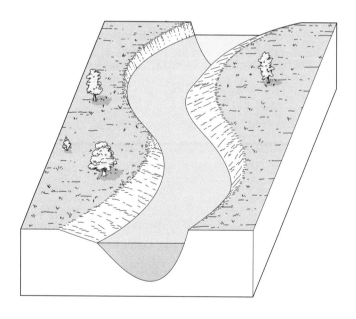

4.   Label the type of drainage patterns illustrated by each of the block diagrams below.

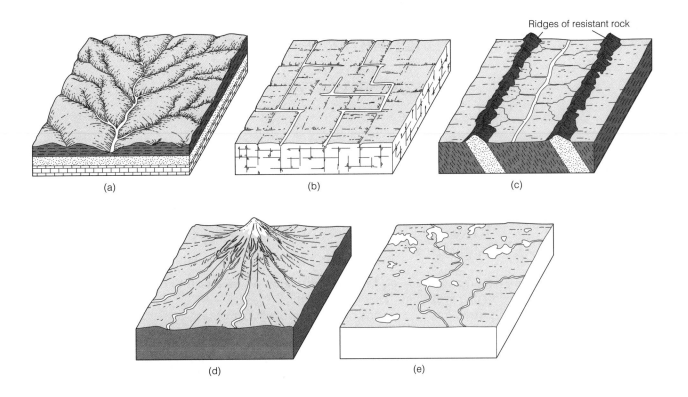

(a)   (b)   (c)

Ridges of resistant rock

(d)   (e)

5.   Identify the feature pointed out in the adjacent photograph. Figure 15.24 in the textbook is a similar photograph.

CHAPTER 16

GROUNDWATER

CHAPTER OBJECTIVES

By the end of this chapter you should be able to:

1. Explain the difference between porosity and permeability and describe factors that influence porosity and permeability of a sediment or sedimentary rock.
2. Describe the relationship between the water table, the capillary fringe, the zone of aeration, and the zone of saturation.
3. Compare groundwater movement in the zone of aeration with that in the zone of saturation.
4. Explain what springs are and what they indicate.
5. Explain what artesian wells are and the condition that result in them.
6. Describe the characteristics and origin of karst terrain.
7. Explain how cave systems develop and how this development is related to the water table.
8. Describe a number of cave formations.
9. Discuss what happens when a well is drilled and some of the consequences of water production in excess of recharge.
10. Describe several sources of groundwater contamination.
11. Differentiate between hot springs and geysers.
12. Discuss how geothermal energy is produced, its merits and problems.

USEFUL ANALOGIES

To illustrate the differences between porosity and permeability look at a block of styrofoam cut open and compare it with a sponge. You can easily see that both have high porosity since each has as much air in its total volume as solid mass. However, if you allow water to drip on both only the sponge will absorb water. In fact, styrofoam is used to make cups! You don't see many sponge cups! Styrofoam, therefore, has low permeability.

A sponge also is good for illustrating how water can move against gravity by capillary pull. If you barely touch a sponge to the surface of a puddle of water it will flow upward. This is what happens to groundwater at the capillary fringe. Capillary action, high porosity and high permeability are what make sponges useful.

KEY TERMS

After reading this chapter, you should be familiar with the following terms.

groundwater
porosity
permeability
aquifer
aquiclude
zone of aeration
suspended water
zone of saturation
capillary fringe
water table
recharge
spring
perched water table
water well
cone of depression
artesian system
artesian-pressure surface
sinkhole
karst topography

solution valley
disappearing streams
cave
cavern
dripstone
stalactites
stalagmites
column
drip curtain
travertine terrace
saltwater incursion
cone of ascension
subsidence
hot springs
geyser
travertine tufa
siliceous sinter
geyserite
geothermal energy

CHAPTER CONCEPT QUESTIONS

1. Why do you think the study of groundwater has emerged as a high priority of geologists in recent years?

2. How do porosity and permeability differ? Which is of greater consideration in the study of groundwater and why?

3. What is the difference between an aquifer and an aquiclude? What characteristics of a rock formation would cause it to be one or the other? What types of rocks make good aquifers and good aquicludes?

4. How can normally non-porous rocks such as basalt or most limestone be modified so that they become effective aquifers?

5. Compare and contrast the movement of groundwater in the zone of aeration, the zone of saturation, and the capillary fringe. What drives the water movement in each?

6. What do springs indicate? Why do springs sometimes occur at different elevations, even in a small area?

7. What causes a producing water well to go dry?

8. What is meant when a well is described as being artesian? What geologic conditions cause an artesian well?

9. What is meant when an areas topography is described as being karst? What geologic conditions cause karst topography?

10. Describe two ways in which sinkholes form?

11. Describe the evolution of a cave and its speleothems (the stalactites, stalagmites, etc.).

12. What is the High Plains aquifer? Why is it important and how is it endangered?

13. What geologic conditions make an area susceptible to saltwater incursion? What human actions cause it to occur?

14. What causes subsidence to occur as a result of groundwater pumping? Give two examples of where this has occurred.

15. List four sources of ground water pollution.

16. Where and how do geysers and hot springs form? What conditions cause a hot spring to be a geyser?

17. How can hot groundwater be harnessed for energy? What are some advantages and disadvantages of this energy source? Name two places in the world where geothermal energy is being used today.

COMPLETION QUESTIONS

1. _____ is the percentage of space in a body of sediment or rock and _____ is the ability of that material to transmit fluid.

2. Groundwater is transmitted in a permeable layer called an _____ and kept in that layer by surrounding impermeable layers called _____.

3. For a material to be permeable it must have _____ pore spaces.

4. The _____ separates the zone of aeration from the zone of saturation.

5. At the water table is a very narrow zone called the _____ where the water is transmitted upward.

6. In the zone of aeration groundwater flows downward driven by _____.

7. In the zone of saturation groundwater flows laterally from regions of _____ pressure to regions of _____ pressure.

8. A spring forms where the _____ intersects the surface of the ground.

9. A water well is an artificial opening into the zone of _____.

10. A spring may form above the regional water table if the ground surface intersects a _____ aquifer.

11. To be effective a water well should be drilled to the zone of _____.

12. Pumping of groundwater creates a cone of _____ around the well.

13. For artesian conditions to exist an aquifer must be confined above and below by _____, the rock sequence must be _____ so that the aquifer formation is exposed at a higher elevation in the recharge area, and there must be sufficient _____ in the recharge area to keep the aquifer filled.

14. In an artesian well, pressure pushes the water up toward the _____ surface.

15. Karst topography develops in a terrain underlain by _____ and is due to dissolution by _____ acid in groundwater.

16. A _____ results when a cave collapses.

17. In a cave _____ form on the ceiling and _____ form on the floor, and if they join they form a _____.

18. The High Plains aquifer accounts for approximately ___ percent of the water used for irrigation in the U.S.

19. In coastal areas undergoing rapid development, removal of excessive amounts of fresh groundwater is resulting in _____.

20. Reduction in water pressure in the _____ spaces of sediments is resulting in compaction of the sediments and subsidence in large areas of the arid western U.S.

21. Careless handling of sewage, landfills, and toxic waste disposal can result in _____ of the groundwater.

22. To be considered a hot spring the temperature of the water must exceed ___ degrees C.

23. The heat for most hot springs and geysers comes from cooling _____ beneath the surface.

24. Deposits of _____ often collect around the vents of geysers.

25. In a geyser the water is forced out of the ground by expanding _____.

26. In some areas of the world like California, Iceland, and New Zealand, _____ energy is used to generate electricity or heat homes.

MULTIPLE CHOICE

1. Groundwater occurs in
   a. the pore spaces between clasts of sand or gravel.
   b. in fractures.
   c. aquifers.
   d. all of the above

2. Porosity is
   a. the ability to transmit fluids.
   b. the amount of void space between grains.
   c. the amount of water pressure present.
   d. none of the above

3. Permeability is
    a. the ability to transmit fluids.
    b. the amount of void space between grains.
    c. the amount of water pressure present.
    d. none of the above

4. The water table marks the
    a. top of the zone of aeration.
    b. the bottom of the zone of saturation.
    c. the top of the zone of saturation.
    d. none of the above

5. The capillary fringe is located
    a. at the water table.
    b. below the zone of saturation.
    c. above the zone of aeration.
    d. none of the above

6. In the zone of saturation groundwater moves
    a. from regions of high pressure to regions of low pressure.
    b. downward toward a lake or stream.
    c. both a and b
    d. neither a nor b

7. A spring is
    a. a man made opening in the zone of saturation.
    b. found where the zone of saturation intersects the ground surface.
    c. where the zone of aeration meets the ground surface.
    d. none of the above

8. A spring located at an elevation above the water table probably indicates
    a. excessive production of wells in the area.
    b. the presence of a perched aquifer.
    c. collapsing caves below the ground.
    d. a problem with saltwater incursion.

9. When water is pumped out of the ground, it creates a _____ around the well.
    a. cone of ascension
    b. cone of depression
    c. cone of incursion
    d. cone of recharge

10. Artesian conditions require
    a. confinement of the aquifer between aquicludes.
    b. a recharge area lower in elevation than the area being produced.
    c. arid conditions in the recharge area.
    d. all of the above

11. In an artesian system the artesian-pressure surface is always located
    a. above the producing aquifer.
    b. below the producing aquifer.
    c. above the ground.
    d. at sea level.

12. Karst topography results from
    a. the incursion of salt water in coastal areas.
    b. the existence of hundreds of artesian wells along a mountain front.
    c. the dissolution of limestone bedrock by groundwater.
    d. all of the above

13. Stalactites and stalagmites form when a cave is
    a. just beginning to dissolve.
    b. in the zone of aeration.
    c. in the zone of saturation.
    d. beneath the zone of saturation.

14. A vertical sheet of rock hanging from a fracture in the ceiling of a cave is
    a. a stalactite.
    b. a stalagmite.
    c. a travertine terrace.
    d. a drip curtain.

15. The effects of salt water incursion has been effectively combated by
    a. pumping fresh water back into the aquifer.
    b. increasing the rate of pumping of the producing wells.
    c. pumping "salt-cleaning" chemicals into the aquifer.
    d. All of the above have been successful.

16. Over pumping of groundwater causes
    a. lowering of the water table.
    b. saltwater incursion.
    c. ground subsidence.
    d. all of the above

17. Which of the following is the best place for a septic tank?
    a. in the zone of aeration in material of high permeability
    b. in the zone of aeration in material of low permeability
    c. in the zone of saturation in material of high permeability
    d. in the zone of saturation in material of low permeability

18. The temperature of a hot spring is
    a. above the boiling point.
    b. above that of the human body.
    c. above the average air temperature of a region.
    d. above 10 degrees C.

19. Siliceous material deposited around the vent of a geyser is called
    a. travertine terrace.
    b. travertine tufa.
    c. sinter.
    d. none of the above

20. Geothermal energy makes up about ___ % of the energy harnessed by humans.
    a. 2
    b. 10
    c. 50
    d. 75

TRUE OR FALSE

___1.   Groundwater commonly occurs as underground rivers.

___2.   Porosity is determined by the size, shape, and arrangement of the grains, by their sorting, and the degree of cement present.

___3.   If a material is porous, it is also permeable.

___4.   In the zone of saturation the pore spaces are completely filled with water.

___5.   The water table approximately follows the overlying surface topography.

___6.   The water well is an artificial opening in the zone of saturation.

___7.   To have a water well pump water all year long, the well must be in the zone of aeration all year long.

___8.   Artesian systems are highly desirable because they cannot be overproduced.

___9.   Low concentrations of sulfuric acid in rain water are responsible for most cave formation.

___10.   Sinkholes are common features of Florida limestone.

___11.   Sinkholes form by an explosive eruption of carbon dioxide gas released by the dissolving of limestones.

___12.   Dripstone deposits are composed of calcium carbonate.

___13.   Because salt water is so much less dense than fresh water, salt water incursion in a coastal aquifers can be mitigated by deepening a well.

___14.   Some of America's most productive agricultural land is in danger because of overproduction of groundwater.

___15.   Over 50% of the water currently used in the U.S. is groundwater.

___16.   Unlike surface water, groundwater is generally impervious to industrial pollution.

___17.   Hot springs are most abundant in the far western United States.

___18.   Geothermal water can be very corrosive.

___19.   Geysers form as a solidified magma forces groundwater to the surface.

___20.   A primary attraction of geothermal energy is that it cannot be exhausted.

DRAWINGS AND DIAGRAMS

1. On the cross-section face of the block diagram below, draw a line representing the water table. Label the zones of saturation and aeration.

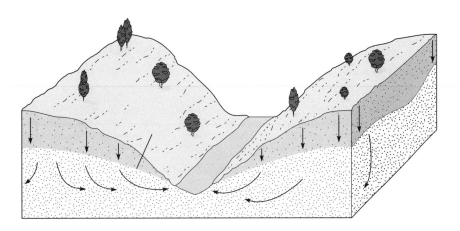

2. Sketch the interior of a cave with stalagmites, stalactites, soda straws and drip curtains. You can check your sketch against Figure 16.17.

CHAPTER 17

GLACIERS AND GLACIATION

CHAPTER OBJECTIVES

By the end of this chapter you should be able to:

1. Distinguish between a glacier and other masses of snow and ice.
2. Describe how glacial ice originates and how it moves.
3. Distinguish between a valley glacier and a continental glacier.
4. Explain the concept of glacial budget and what determines if a glacier will advance or retreat.
5. Describe how rates of glacial movement varies within a glacier and with the seasons.
6. List and describe the various erosional processes associated with glaciers.
7. List, describe and recognize the various erosional and depositional landforms associated with both continental and valley glaciers.
8. Describe how the climate was different in the world during the Pleistocene than it is today and what effects that different climate had on the landscape in different areas, particularly the arid western U.S. and coastal areas.
9. Discuss several theories for the cause of the Pleistocene Ice Ages.

USEFUL ANALOGIES

Have you ever left an ice tray in the freezer for a long period of time, then wondered why the cubes were so small when you finally wanted some ice? That's sublimation!

Lots of solids flow plastically given enough time. Old window glass panes often will be wavy and even may be thin at the top and thick at the bottom. This is because they flow, even more slowly than glacial ice.

KEY TERMS

After reading this chapter, you should be familiar with the following terms:

glacier   sublimation
calving   firn

glacial ice
plastic flow
basal slip
valley glacier
continental glacier
ice cap
glacial budget
zone of accumulation
zone of wastage
firn limit
terminus
stagnant glacier
crevasse
ice fall
glacial surge
outlet glacier
glacial erratics
plucking
roche moutonnée
abrasion
glacial polish
glacial striations
rock flour
U-shaped glacial trough
truncated spurs
fiord
hanging valley

cirque
tarn
arête
horn
ice-scoured plain
glacial drift
till
stratified drift
end moraine
ground moraine
recessional moraine
terminal moraine
lateral moraine
medial moraine
drumlin
outwash plain
valley train
kettle
kame
esker
varve
dropstone
Pleistocene
pluvial lake
proglacial lake
Milankovitch Theory

CHAPTER CONCEPT QUESTIONS

1. What is the difference between a glacier and a simple patch of mountain snow or ice? Why is an iceberg not a type of glacier?

2. How does glacial ice form? What makes it different from other ice, for example that on a frozen lake?

3. What causes a glacier to move? Describe the two mechanisms by which glaciers move.

4. Compare and contrast a valley glacier with a continental glacier.

5. Explain in terms of glacial budget why a glacier advances or retreats.

6. How do flow rates vary within a single glacier.

7. Where, how and why do crevasses form?

8. What factors influence the rate at which a glacier will flow?

9. What is the difference between plucking and abrasion. What are the results of each?

10. How do you tell a valley that has been cut by a stream from one which has been carved by a glacier?

11. Explain the origin of the following: (a) fiord, (b) hanging valley, (c) arête, (d) cirque, (e) horn.

12. Describe a landscape, such as that of Canada, that has been subjected to the scouring of a large continental glacier.

13. Contrast the origin and characteristics of stratified drift and till.

14. Compare and contrast recessional moraines, terminal moraines, lateral moraines and medial moraines. Which are restricted to valley glaciers?

15. Compare and contrast the origin of outwash plains, valley trains, kames and eskers.

16. Contrast the origin of the following types of glacial lakes: (a) tarns, (b) kettles, (c) pluvial lakes, (d) proglacial lakes. Characterize the sedimentary deposit of a glacial lake.

17. Certain regions of Earth experienced notable differences in climate during the Pleistocene compared to today, even outside the polar regions. Describe some of these.

18. How did the climate during the Pleistocene differ for the Earth as a whole? More specifically, how did it differ for the Sahara Desert and the arid western U.S. How did the glaciation effect the coastlines of the continents?

19. What is meant by isostatic rebound? How does it relate to Pleistocene glaciation? How and why has it varied in different parts of the North American Continent?

20. Describe the three parameters that vary to cause glaciations according to the Milankovitch theory.

21. What theories have been proposed for causes of short-term cold periods such as the Little Ice Age?

COMPLETION QUESTIONS

1. A glacier is a mass of ice composed of compacted and recrystallized _____ that flows under its own weight across the land.

2. Today glaciers cover about _____ of the surface of the land, they occur on all continents except _____.

3. The largest present day glaciers occur in _____ and _____.

4. About _____ % of the worlds water is contained in glaciers.

5. Icebergs are masses of floating ice that enter the sea by breaking off a glacier in a process called _____.

6. Ice can evaporate without melting in a process called _____.

7. Because ice is a crystalline solid with specific chemical and physical properties, it can be classified as a _____.

8. As snow accumulates, it is _____ by the weight of the overlying snow, refreezes and forms a granular type of ice called _____ which further compacts to form glacial ice.

9. At a critical thickness of _____ meters, ice begins to deform and flow.

10. Once a mass of recrystallized ice begins to move, it then can properly called a _____.

11. _____ flow, permanent deformation of the ice, is the most important way that glaciers move.

12. Meltwater can reduce the _____ between the ice and the ground beneath allowing movement by the process of _____.

13. Of the two main types of glaciers, _____ glaciers are the largest, covering an area greater than _____ km^2 .

14. _____ glaciers are confined to high mountain valleys.

15. In Greenland and Antarctica continental glaciers are more than _____ meters thick.

16. A glacier has two major zones. Yearly snow additions exceed losses in the zone of _____ but losses exceed additions in the zone of _____. The _____ limit separates the two.

17. Accumulation is due to _____ fall, and wastage is due to _____, _____, and the calving of _____.

18. If wastage exceeds accumulation, a glacier will _____, but if accumulation exceeds wastage, a glacier will _____; and if wastage and accumulation are in balance, the glacier will be _____.

19. In a valley glacier, the highest velocity is in the zone of _____.

20. During the winter a glacial moves only by _____ flow, but during the warmer months, when meltwater is abundant, _____ is also an important mechanism.

21. The upper part of a glacier behaves as a _____ solid, and this causes _____ to develop when a glacier moves over irregularities in the valley floor.

22. During a glacial _____ glaciers may advance several kilometers in a year or less.

23. Flow rates in continental glaciers are measured in _____ to _____ per day.

24. Glacial _____ are large boulders transported many tens to hundreds of kilometers from their points of origin.

25. _____ occurs when glacial ice freezes in the cracks and crevices of a bedrock projection and pulls it loose. This results in a bedrock landform called a _____.

26. Abrasion by glaciers results in a smooth glacial _____ which is often marred by straight, shallow scratches called _____ .

27. The milky appearance of streams containing meltwater is generally due to rock _____.

28. Glacially eroded valleys are characterized by having a _____ shaped cross section.

29. As glaciers melt and sea level rises, glacial troughs may be flooded by the sea resulting in _____ .

30. Small tributary glaciers that join a larger valley glacier produce _____ valleys.

31. A bowl-shaped depression at the head of a glacier is called a _____, and the knife-like ridge separating two of these is called a _____.

32. A sharp mountain peak surrounded on all sides by cirques is a _____.

33. Areas eroded by _____ glaciers tend to be rather flat, monotonous and characterized by a _____ drainage pattern.

34. In North America, a large part of the _____ shield was stripped of its soil and unconsolidated surface sediment revealing extensive exposures of polished and striated bedrock.

35. All sediment deposited by a glacier is called _____.

36. If glacial sediment was deposited directly by ice leaving a poorly sorted disorganized deposit it is called _____. If however, the deposited was sorted by streams it is _____ drift.

37. A ridge of till deposited at a glacier's terminus is called an _____ moraine and one deposited at the side of a valley glacier is a _____ moraine.

38. If an end moraine marks the furthest advance of a glacier it is a _____ moraine.

39. If an end moraine is deposited by a minor glacial advance during an overall recession it is a _____ moraine.

40. Where two lateral moraines merge, a _____ moraine forms.

41. When a continental glacier overrides ground moraine it often shapes the till into elongate hills called _____.

42. Meltwater laden with sediment from a melting glacier forms deposits called _____ plains and _____ trains.

43. Depressions that form from a block of ice left in the outwash produces a _____.

44. Long sinuous ridges of stratified drift deposited by streams flowing under a glacier are called _____.

45. Stratified drift deposited in depressions in the surface of the ice and subsequently lowered to the ground when the ice melts are called _____.

46. Glacial lake sediments with alternating light and colored laminae are called _____.

47. In terms of time, one light and one dark couplet of a varve equals one _____.

48. Large stones deposited in otherwise very fine-grained lake sediment are called _____ and were probably carried into the lake by _____.

49. The Pleistocene glaciation began about _____ million years ago.

50. _____ major ice advances have been recognized in North America while Europe had six or seven. However at least _____ warm-cold cycles can be detected in deep sea cores.

51. The Pleistocene glaciation affected the _____ hemisphere.

52. The Great Salt Lake is remnant of a once larger _____ lake called Lake Bonneville.

53. Proglacial lakes are impounded by the _____ one side and a _____ on the other.

54. The Great lakes presently contain _____% of the water contained in the Earth's freshwater lakes.

55. When the Pleistocene glaciation was at its maximum, sea level was _____ meters lower than it is today.

56. If the present glaciers melted, sea level would _____ by _____ meters.

57. Under the weight of the Pleistocene glacial ice, the crust of the Earth was depressed up to _____ meters below its position today.

58. Since the Pleistocene continental glaciers have melted, the land has been rising. This process is called glacial _____.

59. According to the _____ theory, glacial-interglacial episodes of the Pleistocene may be due to variations in the eccentricity of the Earth's orbit, the tilt of the axis and the precession of the equinoxes.

MULTIPLE CHOICE

1. Which of the following is NOT a necessary condition for a mass of ice to be considered a glacier?
   a. It must advance in the winter more than it melts in the summer.
   b. It must originate by compaction and recrystallization of snow.
   c. It must flow under its own weight.
   d. It must be on land.

2. During the transformation from snow to glacial ice the percentage of air
   a. decreases.
   b. increases.
   c. is unaffected.

3. When a glacier sublimates it
   a. melts.
   b. evaporates without melting.
   c. travels forward faster than normally.
   d. forms icebergs.

4. The critical depth of snow and ice needed to cause the ice below to deform and flow is
   a. 4 meters.
   b. 40 meters.
   c. 400 meters.
   d. 4 kilometers.

5. During the Pleistocene, continental glaciers covered large parts of
   a. North America and South America.
   b. North America, South America and Europe.
   c. North America, Europe and Asia.
   d. North America, Asia and Africa.

6. When wastage exceeds accumulation, a glacier will
    a. retreat.
    b. advance.
    c. remain stationary.
    d. stop flowing.

7. During warmer periods velocity due to basal slip _____ and velocity due to plastic flow _____.
    a. decreases / increases.
    b. increases / decreases.
    c. increases / stays the same.
    d. stays the same / decreases.

8. Crevasses result from the _____ behavior of ice in the top forty meters of a glacier.
    a. brittle
    b. plastic
    c. elastic
    d. juvenile

9. Glacial abrasion produces
    a. rock flour.
    b. glacial polish.
    c. glacial striations.
    d. all of these

10. Which is NOT characteristic of a glacial trough?
    a. a broad, flat valley floor
    b. steep, often vertical valley walls
    c. V-shaped cross-section
    d. truncated spurs

11. Which of the following pair are both erosional in origin?
    a. cirques and moraines
    b. horns and cirques
    c. cirques and drumlins
    d. moraines and drumlins

12. Glacial till is
    a. unstratified drift.
    b. stratified drift.
    c. both a and b
    d. neither a nor b

13. Medial moraines form when
    a. two terminal moraines merge.
    b. two lateral moraines merge.
    c. an end moraine is over-ridden by a glacier.
    d. a mountain peak is surrounded on all sides by glaciers.

14. Which of the following is likely to show poor sorting and no stratification?
    a. an esker
    b. an outwash plain
    c. a drumlin
    d. a kame

15. Eskers form
    a. on top of the ice.
    b. in front of the ice.
    c. beneath the ice.
    d. none of the above

16. A recessional moraine is a type of
    a. end moraine.
    b. lateral moraine.
    c. medial moraine.
    d. ground moraine.

17. During periods of glacial advances the Sahara Desert was covered by
    a. glaciers.
    b. forests.
    c. a shallow sea.
    d. volcanoes.

18. The Great Lakes started out as
    a. kettles.
    b. proglacial lakes.
    c. pluvial lakes.
    d. tarns.

19. Pluvial lakes formed in the far western U.S. during the Pleistocene glaciation as a result of
    a. blocks of ice melting in outwash plains.
    b. impoundment of water by terminal moraines.
    c. water filling depressions carved in bedrock by glaciers.
    d. an overall cooler and wetter climate in the area.

20. When large continental glaciers advanced across the northern hemisphere during the Pleistocene
    a. the crust beneath the subsided.
    b. worldwide sea level fell.
    c. temperate, subtropical and tropical belts were compressed toward the equator.
    d. all of the above

21. The best explanation for short periods of glacial advance that span only a few centuries includes
    a. plate tectonics.
    b. Milankovitch cycles.
    c. frequent volcanic eruptions coupled with variations in solar energy.
    d. all of the above

TRUE OR FALSE

___1.   The most common type of glacier is frozen sea ice.

___2.   Ice is a mineral.

___3.   Africa is the only continent which lacks glaciers.

___4.   In the western U.S. glaciers are an important source of fresh water.

___5.   Glaciers are made of firn.

___6.   As you travel north you can expect to encounter glaciers at progressively lower elevations.

___7.   During most of the year basal slip is the main mechanism by which glaciers move.

___8.   Continental glaciers are unconfined by topography.

___9.   The main valley glacier moves faster than its tributaries.

___10.   Glacial striations are parallel to the direction of glacial flow.

___11.   Glacial troughs and stream cut valleys cannot be distinguished by shape of profile alone.

___12.   During the glacial advances of the Pleistocene Epoch, sea level was higher than it is today.

___13.   Fiords are glacial troughs occupied by the sea.

___14.   Horns are deposited by meltwater beneath a glacier.

___15.   Arêtes are bowl-shaped depressions near the head of a glacier.

___16.   Stratified drift is called till.

___17.   Medial moraines form at the end of a valley glacier.

___18.   Drumlins can be used to determine the direction of movement of a continental glacier.

___19.   Most outwash streams are meandering rivers.

___20.   The dark layer of a varve is deposited in the winter.

___21.   The discovery of widespread glaciers during the Pleistocene was first suggested by the scientists of ancient Greece and Rome.

___22.   During the Pleistocene, the world's temperature was close to freezing.

___23.   Since the last glacial advance the northern third of North America has rebounded considerably more than the southern two thirds.

___24.   Plate tectonics may play a part in causing glaciations by moving continents to polar areas.

DRAWINGS AND FIGURES

1. On the block diagram below, label (a) a hanging valley, (b) a horn, (c) an arête, (d) a U-shaped glacial trough, (e) a cirque, and (f) truncated spurs.

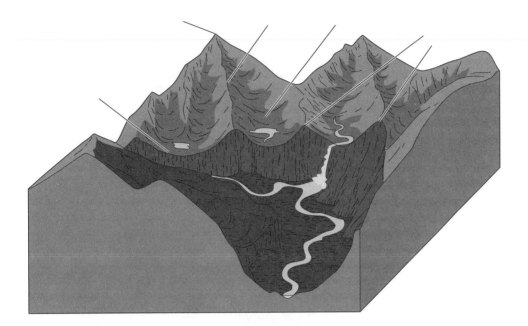

2. On the block diagram below label a lateral moraine, medial moraine, and the ground moraine

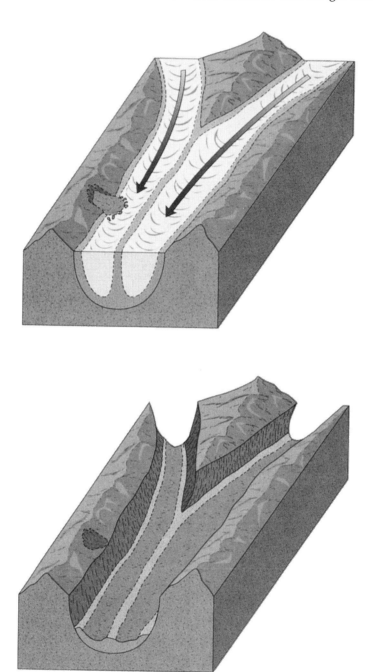

3. On the block diagram below, label (a) a kame, (b) drumlins, (c) the outwash plain, (d) kettle lakes, (e) an esker and (f) the end moraine.

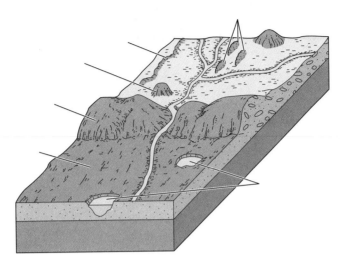

CHAPTER 18

THE WORK OF WIND AND DESERTS

CHAPTER OBJECTIVES

By the end of this chapter you should be able to:

1. Describe how clasts of various sizes are moved by wind.
2. Compare and contrast water and wind as transporting agents.
3. Describe the main processes and products of wind erosion.
4. Define deflation and describe some of the products of that process.
5. Describe how dunes form and migrate.
6. Classify and describe the major dune types.
7. Define loess, and discuss its origin, character and distribution.
8. Describe what factors determine the distribution of deserts.
9. Describe the climatic conditions that constitute a desert and discuss how desert vegetation has adapted to these conditions.
10. Describe the important weathering processes of deserts and their products.
11. Describe how the surface and groundwater conditions in deserts differs from other areas.
12. Characterize, recognize and discuss the origin of desert landforms.

KEY TERMS

After reading this chapter , you should be familiar with the following terms:

suspended load
bed load
saltation
abrasion
ventifact
yardang
deflation
deflation hollow
desert pavement
dune
wind shadow

barchan dunes
longitudinal dunes
transverse dunes
barchanoid dunes
parabolic dunes
loess
Coriolis effect
desert
rainshadow desert
rock varnish
internal drainage

playa lakes   pediment
playa or salt pan   inselberg
alluvial fan   mesa
bajada   butte

CHAPTER CONCEPT QUESTIONS

1. Distinguish between the bed load and the suspension load. What sizes are included in each in wind transport? Compare this to transport by water.

2. Why are sand-sized particles generally entrained by wind before finer silt and clay?

3. How does wind abrade rocks? Describe two products of abrasion commonly found in deserts.

4. What is deflation? Describe two products of deflation commonly found in deserts.

5. How do sand dunes form? How do they migrate?

6. What are the main factors which influence the size, shape, and arrangement of dunes?

7. List the four major dune types. For each describe (a) the orientation relative to the prevailing wind, (b) the conditions (i.e. sand supply, vegetation) under which that type forms, and (c) the typical range of sizes encountered.

8. What is loess? Where does it come from? Why is it important to agriculture?

9. What is the Coriolis effect and what causes it?

10. What is the difference between a semi-arid and an arid region?

11. What is the difference between a low to mid-latitude desert and a rainshadow desert?

12. Characterize a desert in terms of temperature range, amount and nature of precipitation, and soil development.

13. List at least two reasons why water erosion is so prominent in deserts despite the low rainfall.

14. How does the drainage pattern in deserts differ from that in more humid areas.

15. What is the source of water in large desert rivers such as the Nile, the Rio Grande and the Colorado? How does the relationship between groundwater and these rivers differ from the relationship of groundwater to surface water in more humid areas (see chapter 16)?

16. What are the major characteristics of a playa lake and salt pan?

17. Define and describe the origin of the following desert landforms: (a) alluvial fan, (b) bajada, (c) pediment, (d) inselberg, (e) mesa, and (f) butte.

COMPLETION QUESTIONS

1. Wind transports fine-grained sediments in _____, and sand-particles by _____.

2. In a desert, the most important erosional agent is _____.

3. Abrasion causes a sandblasting effect to form wind-faceted rocks called _____.

4. The removal of loose sediment by the wind is called _____.

5. Deflation may leave a blown out depression in the surface called a _____ or a mosaic of close-fitting pebbles left behind called _____.

6. Sand dunes form when sand accumulates in the wind _____ of an obstacle.

7. Most sand dunes have an asymmetric profile with a gentle _____ slope and a steep _____ slope.

8. The steepest slope of a dry sand dune is between ____ and ____ degrees, the angle of _____.

9. A crescent-shaped dune with its tips pointing downwind is called a _____ dune, while a crescent-shaped dune with its horn pointing upwind is called a _____ dune.

10. A long, linear dune perpendicular to the wind is called a _____ dune, and a long, linear dune parallel to the prevailing wind direction is called a _____ dune.

11. _____ is fine-grained, wind transported material derived from deserts, glacial outwash, and floodplains in semi-arid areas.

12. Loess covers _____ percent of the Earth's land surface and _____ percent of the land surface in the United States.

13. Deserts receive less than _____ centimeters of rainfall per year.

14. Because of the _____ effect, winds are deflected to the _____ of their direction of motion in the northern hemisphere and to the _____ in the southern hemisphere.

15. Air that rises at the equator tends to cool and sink in belts between ____ and ____ degrees north and south latitude.

16. The Great Basin Region of the western U.S. is largely a vast _____ desert formed on the _____ side of the Sierra Nevada mountain range.

17. _____ weathering dominates in a desert, the main processes being _____ fluctuations and _____ wedging.

18. _____ is a coating composed of iron and manganese _____ and _____.

19. Desert valleys in the southwestern United States commonly have _____ drainage resulting in the formation of _____ lakes.

20. Alluvial _____ form at the mouth of _____ where streams _____ their load.

21. When two or more alluvial fans coalesce, a _____ can form.

22. Sloping surfaces extending from the base of a mountain are called _____.

23. Isolated remnants of mountains in desert areas are called _____.

24. _____ are broad, flat topped erosional remnants bounded on all sides by steep slopes. A smaller version of this is called a _____.

MULTIPLE CHOICE

1. Wind-blown sand is generally transported by
    a. saltation.
    b. suspension.
    c. either a or b
    d. neither a nor b

2. Wind generally entrains silt and clay-sized grains _____ sand-sized grains.
    a. before
    b. after
    c. about the same time as

3. In a desert, wind erosion is _____ erosion by running water.
    a. less important than
    b. more important than
    c. about equally important as

4. Abrasion by wind generally takes the form of
    a. pitting.
    b. etching.
    c. polishing.
    d. all of the above

5. Deflation can only occur if
    a. the sediments are loose.
    b. the sediments are anchored by vegetation.
    c. if the sediments are wet.
    d. all of the above

6. The steep side of an asymmetric sand dune
    a. faces upwind.
    b. faces downwind.
    c. can face either upwind or downwind.
    d. can not be used to determine the wind direction.

7. Sand dunes develop where sand accumulates in a
    a. rainshadow.
    b. deflation hollow.
    c. wind shadow.
    d. none of these

8. Barchan dunes
    a. are crescent-shaped with tips pointing downwind.
    b. are crescent-shaped with tips pointing upwind.
    c. are linear dunes oriented parallel to the prevailing wind.
    d. are linear dunes oriented perpendicular to the prevailing wind.

9. Transverse dunes
   a. are crescent-shaped with tips pointing downwind.
   b. are crescent-shaped with tips pointing upwind.
   c. are linear dunes oriented parallel to the prevailing wind.
   d. are linear dunes oriented perpendicular to the prevailing wind.

10. Ancient sand dunes, now preserved as sandstones, would be characterized by an abundance of which sedimentary structure?
    a. scour and fill
    b. mud cracks
    c. graded bedding
    d. cross-stratification

11. Loess is dominated by
    a. silt and clay.
    b. sand and silt.
    c. sand and gravel.
    d. gravel.

12. The fertile soils of the Great Plains, the Midwest, and in the Mississippi Valley were derived from extensive
    a. sand dunes.
    b. loess deposits.
    c. playa lakes.
    d. alluvial fans.

13. In high pressure zones
    a. hot, moist air rises.
    b. hot, moist air descends.
    c. dry, cool air rises.
    d. dry, cool air descends.

14. Deserts receive less than _____ centimeters of rain per year.
    a. 15
    b. 2.5
    c. 35
    d. 25

15. Most of the world's deserts are found in belts
    a. along the equator.
    b. in the low to middle latitudes.
    c. in the middle to high latitudes.
    d. in the polar regions.

16. Where a high mountain range causes a rainshadow desert, the desert forms on _____ side(s) of the mountains.
    a. upwind (windward)
    b. downwind (leeward)
    c. both

17. In most desert regions, the water table is
    a. higher than the stream channels.
    b. below the stream channels.
    c. equal to the position of the stream channels.
    d. none of the above.

18. Rock varnish contains oxides of _____ and iron.
    a. copper
    b. tin
    c. magnesium
    d. none of the above

19. Most deserts are dominated by a surface consisting of
    a. a cover of sand dunes.
    b. exposures of bedrock and desert pavement.
    c. loess deposits.
    d. playa lakes.

20. Alluvial fans coalesce to form
    a. playas
    b. pediments
    c. bajadas
    d. buttes

TRUE OR FALSE

___1.   Wind is a turbulent fluid.

___2.   Sand is commonly lifted up to ten meters by wind.

___3.   In saltation, sand grains move in a bouncing fashion.

___4.   The suspension load of wind may be carried thousands of kilometers.

___5.   In a desert, running water is the most important erosional agent.

___6.   Ventifacts form by deflation.

___7.   Barchan dunes have their steep side facing downwind, and their tips pointing upwind.

___8.   Parabolic dunes form where there is a partial cover of vegetation.

___9.   Longitudinal dunes require large sediment supplies to form.

___10. Sand usually travels by saltation up the leeward side of a dune, then avalanches down the windward side.

___11. Accumulations of loess deposits are a common feature of deserts.

___12. Loess deposits can cause big problems for agriculture because of their low fertility.

___13. In the Southern Hemisphere, the Coriolis effect deflects objects to the left of their intended path.

___14.   Deserts form in belts of high pressure at the low to middle latitudes.

___15.   Deserts have high temperatures throughout the year.

___16.   By definition deserts have no plants living in them.

___17.   Evaporite minerals may be common in playa lakes.

___18.   It is not uncommon for some deserts to receive their full year's rainfall in one cloudburst.

___19.   Buttes and mesas form by streams eroding through a resistant cap rock to erode the less resistant rock below.

___20.   Pediments are small erosional remnants of mountains.

FIGURES AND DRAWINGS

1.   Identify the type of dunes illustrated by the four block diagrams below.

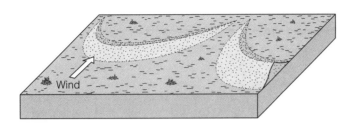

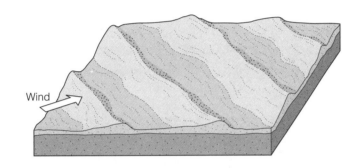

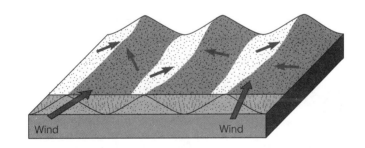

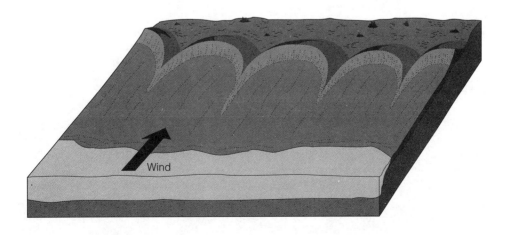

2.   Identify the features illustrated by the following photographs. Hint for photograph a): The vast white area is covered in salt. This photograph is not in book so the answer is given in the back of this manual.

CHAPTER 19

SHORELINES AND
SHORELINE PROCESSES

CHAPTER OBJECTIVES

By the end of this chapter you should be able to:

1. Name the parts of a wave and describe wave motion.
2. Discuss the factors which influence size of waves and those which cause a wave to break.
3. Distinguish between rip currents and nearshore currents and discuss their causes.
4. List and distinguish between the various parts of a beach.
5. Discuss the material, origin, and transportation of beach sediment.
6. Describe the differences between a "summer beach" and a "winter beach" and account for the differences.
7. Define and compare and contrast the origins of spits, baymouth bars and tombolos.
8. Describe a barrier island and list two theories for their origin.
9. Discuss the concept of a shoreline's sediment budget and list the sources of sediment, the processes responsible for sediment removal, and the consequences of negative vs. positive budget.
10. List the main processes responsible for erosion along a shoreline and the landforms that result.
11. Contrast depositional coasts with erosional coasts and emergent coasts with submergent coasts.
12. Discuss the origin of tides and their variation with respect to the relative positions of the sun and moon.

KEY TERMS

After reading this chapter, you should be familiar with the following terms

shorelines	trough
tides	wave length
flood tide	wave height
ebb tide	celerity
spring tide	wave period
neap tide	wave base
crest	seas (in the wave sense)

rogue waves
swells
fetch
breaker
nearshore zone
breaker zone
surf zone
wave refraction
longshore currents
rip currents
beach
pocket beach
backshore
berms
foreshore
beach face
longshore drift
groins
summer beach
winter beach

spit
baymouth bar
tombolo
jetties
barrier island
nearshore sediment budget
corrosion
hydraulic action
abrasion
sea cliffs
wave-cut platform
wave-built platform
headlands
sea cave
sea arch
sea stacks
submergent coast
estuary
emergent coast
marine terrace

CHAPTER CONCEPT QUESTIONS

1. What causes ocean tides? What unique conditions cause spring and neap tides?

2. Define wave length, wave period and celerity. How do they relate to each other?

3. Describe the path that a "water particle" follows as waves advance across the water. How does this change with depth? How does it change as the shoreline is approached?

4. How do waves form? What factors control their size?

5. What are rogue waves and what causes them? What problem can they cause?

6. What causes a swell to change to a breaker?

7. How does the slope of the bottom offshore affect the character of a breaker?

8. What is wave refraction and why does it happen?

9. What are longshore currents and how do they form?

10. What are rip currents and how do they form? What determines their location?

11. Explain the spatial relationships between the following terms: (a) beach, (b) foreshore, (c) backshore, (d) beach face and (e) berms.

12. How do seasons affect the characteristics of a beach?

13. Explain the origin of the following: (a) spit, (b) baymouth bar, and (c) tombolo.

14. What are barrier islands and where are they found? Describe two models for their origin.

15. Explain the concept of a sediment budget for shorelines. What are the sources of sediment supplied to shorelines? What processes deplete shorelines of sediment?

16. List and define the processes which are important in shoreline erosion?

17. Explain how wave-cut platforms develop. How do they evolve into marine terraces?

18. What problem does the worldwide rise in sea level pose for society? What measures are being taken to mitigate the potential damage? What further problems have been caused by some of these measures?

19. What are estuaries and how do they form?

COMPLETION QUESTIONS

1. Shorelines are defined as the areas between _____ tide and the highest level on land effected by _____ waves.

2. A rising tide is called a _____ tide and a falling tide is an _____ tide.

3. The highest tidal ranges occur during the _____ tides, and lower tidal ranges occur during the _____ tides.

4. When the sun and moon are aligned the result is a _____ tide, and when they are at right angles to one another the result is a _____ tide.

5. Though waves have several origins, the majority are generated by the _____.

6. The highest part of a wave is its _____ and the lowest is the _____. The distance between these two points is the wave _____.

7. If two or more waves of different periods happen to have their crests coincide a large _____ can develop.

8. As waves pass the "particles" of water actually follow a _____ path.

9. The depth at which water "particle" motion dies out is called _____, a depth in theory equal to _____ the wave length.

10. As the wind blows across the water beneath a storm center, it forms sharp crested choppy waves called _____.

11. As these waves leave the storm generating area they begin to sort themselves out into _____.

12. A major control of the size to which a wave can develop is the _____, which is the distance the wind has been blowing over the water.

13. As the incoming swell approaches a beach, the wave base intersects the bottom, the waves steepen and become _____.

14. At this point, friction causes the wave velocity to _____, the wave length to _____, and the wave height to _____.

15. As a wave approaches the shore, different parts of the wave travel at different velocities and the wave _____.

16. Longshore currents flow _____ to the shoreline and form when waves arrive at an angle to the beach.

17. _____ currents form by the converging of two longshore currents. They flow toward the _____.

18. A _____ is a deposit of unconsolidated sediment extending landward from low tide to a major change in topography or a line of permanent vegetation.

19. The backshore is that part of a beach only covered by water during extremely high tides or _____ and consists of one or more _____, nearly horizontal platforms deposited by waves.

20. The area of a beach between high and low tide is the _____.

21. Longshore currents transport sand along a beach in a process called longshore _____.

22. Structures built to stabilize the beach by minimizing the effects of longshore currents are called _____.

23. The most common mineral of beach sands is _____.

24. Curving sand bars, fed by longshore currents, are called _____.

25. If a spit grows across an embayment, it forms a _____ bar.

26. A sand bar that connects an island to the mainland is called a _____.

27. Barrier islands are _____ to the shoreline and are separated from the mainland by a _____.

28. Most of the sediment on a beach is supplied by _____ and then dispersed along a beach by _____ currents.

29. Sea cliffs, arches and sea caves form by wave _____.

30. As a sea cliff retreats, a gently sloping platform called a wave-cut _____ forms.

31. Isolated columns of rock left by a receding sea cliff are called _____.

32. If the land goes up with respect to sea level, an _____ coast is formed.

33. If the land goes down with respect to sea level, a _____ coast has formed.

34. When wave-cut platforms are raised above sea level, generally by tectonic processes, they are called marine _____.

MULTIPLE CHOICE

1. The distance from crest to crest is called the
    a. wave period.
    b. wave speed.
    c. wave height.
    d. none of the above

2. In most areas high tides occur
    a. once a day.
    b. twice a day.
    c. four times a day.
    d. once every other day.

3. Spring tidal ranges occur
    a. during full and new moons.
    b. during first and last quarter moons.
    c. during April and May.
    d. none of the above

4. The sequence of wave development is
    a. sea, breaker, swell.
    b. breaker, sea, swell.
    c. sea, swell, breaker.
    d. breaker, swell, sea.

5. If the wavelength is 10 meters, the wave height is 1 meter and the wave period is 50 seconds, calculate the celerity.
    a. 10 meters
    b. 5 seconds/meter
    c. 500 meter seconds
    d. .2 meters per second

6. Waves begin to break when they are in a water depth of _____ times the wave length.
    a. .5
    b. 1.25
    c. 1.0
    d. 1.5

7. The size of a wave is controlled by the
    a. wind speed.
    b. wind duration.
    c. fetch.
    d. all of the above

8. The diameter of the orbitals followed by water "particles" during wave motion ____ with depth.
    a. increase
    b. decrease
    c. are unchanged

9. Swell waves are characterized by having
    a. sharp crests and a variety of sizes.
    b. rounded crests and a consistency of size.
    c. sharp crests and a consistency of size.
    d. rounded crests and a variety of sizes.

10. When a wave refracts it
    a. becomes more parallel to the shoreline.
    b. becomes more perpendicular to the shoreline.
    c. begins to move away from the shoreline.
    d. disappears.

11. The position of rip currents is largely determined by
    a. the strength of the wind
    b. the tidal range
    c. the configuration of the sea floor
    d. the distance to the equator

12. Berms generally compose the
    a. foreshore.
    b. backshore.
    c. beachface.
    d. none of the above

13. The difference between summer and winter beaches is primarily due to
    a. seasonal variations in the strength of tides.
    b. freezing and thawing of ocean water.
    c. seasonal reversals in the direction of longshore currents.
    d. seasonal differences in the frequency and intensity of storms.

14. Spits are
    a. linear bars separated from the land by a lagoon.
    b. curving sand bars fed by rip currents.
    c. connected to an island.
    d. curving sand bars fed by longshore currents.

15. A baymouth bar is
    a. a linear bar separated from the mainland by a lagoon.
    b. a bar that totally seals off an embayment.
    c. a bar that connects an island to the mainland.
    d. none of the above

16. Most sediment is transported to a beach by _____, and distributed along the shoreline by
    _____.
    a. streams / rip currents
    b. longshore currents / rip currents
    c. streams / longshore currents
    d. longshore currents / tides

17. As sea level rises barrier islands
    a. migrate landward.
    b. migrate seaward.
    c. remain fixed.
    d. grow taller.

18. Sediment is removed from shorlines by
    a. wind.
    b. longshore currents.
    c. transport by offshore directed currents.
    d. all of the above

19. As a sea cliff recedes a _____ develop.
    a. tombolo
    b. a wave cut platform
    c. a sea arch
    d. barrier island

20. Which of the following correctly lists the order of formation of landforms as a headland is eroded?
    a. sea cave, sea stack, sea arch
    b. sea stack, sea cave, sea arch
    c. sea cave, sea arch, sea stack
    d. sea arch, sea cave, sea stack

21. Emergent shorelines are due to
    a. falling sea level due to glaciation.
    b. rising land due to local tectonic activity.
    c. rising land due to isostatic rebound.
    d. any of the above

TRUE OR FALSE

___1. Shorelines are attractive to developers because they are geologically more stable than most other areas.

___2. Flood tides occur only during full or new moons and are about 20% higher than other high tides.

___3. Neap tides form when the Sun and the Moon are at right angles to each other with respect to Earth.

___4. As waves move toward the shore in deep water, water "particles" experience little or no net forward movement.

___5. Waves in lakes are generally smaller than in the ocean because the wind cannot blow as hard over land as over the sea.

___6. Wave height is measured from the sea floor to the crest of a wave.

___7. Waves "feel the bottom" when the depth is about .5 times the wave length.

___8. Longshore currents form when waves approach the beach at an angle to the shoreline.

___9. Rip currents flow away from the shoreline, sometimes at high velocities.

___10. The backshore area of a beach is generally covered only during high tide.

___11. Groins have proven nearly 100% effective in halting erosion of beaches.

___12. Typically removal of sand from a beach dominates in the winter and deposition of sand on a beach dominates in the summer.

___13. Winter beaches generally have wider berms than summer beaches.

___14. Jetties may be effective in preventing baymouth bars from blocking harbor entrances.

___15. In the United States the best developed barrier islands are found off the coasts of Washington, Oregon, and northern California.

___16. Erosion of sea cliffs only contributes between 5 and 10 percent of the sediment supplied to shorelines.

___17. Currently sea level, on a worldwide basis, is dropping.

___18. Due to wave refraction, wave energy is concentrated in embayments, and diffused or scattered on headlands.

___19. The presence of sea stacks is indicative of retreating cliffs in an erosion dominated shoreline.

___20. Modern coastal management techniques have proven highly successful in arresting the damage caused by rising sea level.

___21. Estuaries are best developed along submergent coastlines.

DRAWINGS AND FIGURES

1. In the cross-section below, label a wave crest and a trough. Also indicate the wavelength and height. Indicate also the path that a water "particle" takes.

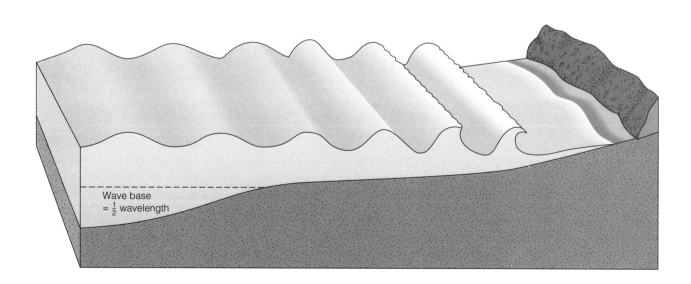

Wave base
= ½ wavelength

2. On the cross-section below label (a) the beach, (b) the beach face, (c) the foreshore, (d) the backshore, and (e) berms.

3. Identify the feature indicated by an arrow in each of the following four photographs.

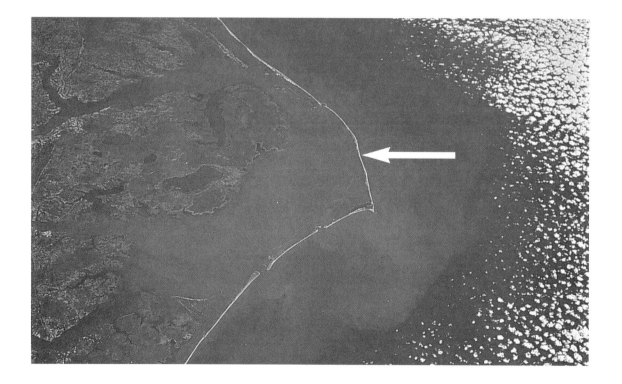

CHAPTER 20

A HISTORY OF THE UNIVERSE, SOLAR SYSTEM, AND THE PLANETS

CHAPTER OBJECTIVES

By the end of this chapter you should be able to:

1. Recount the early history of the universe establishing a chronology for such events as the Big Bang, creation of forces, elements, and the solar system.
2. List the planets in order of distance from the sun and contrast the terrestrial planets and the Jovian planets in terms of size and density.
3. Summarize the solar nebula theory for the formation of the solar system and state why it is the most satisfactory theory to date.
4. Classify and characterize the types of meteorites.
5. Characterize each planet in terms of its internal structure and atmosphere.
6. Explain what is meant when it is said that the Earth is a differentiated planet and explain how differentiation occurs.
7. Describe the internal structure of the moon and contrast the known geology of the lunar maria and highlands.
8. Summarize the most widely accepted theory for the origin of the moon.

KEY TERMS

After reading this chapter, you should be familiar with the following terms.

Big Bang
plane of the ecliptic
gravity
electromagnetic force
strong nuclear force
weak nuclear force
solar system
terrestrial planets
Jovian planets

solar nebula theory
planetesimals
asteroids
comets
meteorites
stones
irons
stony-irons
outgassing

differentiation   (lunar) highlands
maria

CHAPTER CONCEPT QUESTIONS

1.   Develop a time line from the Big Bang to the origin of Earth. Include such events as the Big Bang, the formation of basic forces of Physics, the creation of elements, the formation of stars and galaxies and the formation of Earth and the solar system.

2.   Name and explain the four basic forces of the universe.

3.   Where and how do the elements hydrogen form? How are they dispersed through space?

4.   What are the general characteristics of the solar system which any theory to explain its origin must account for?

5.   What are the major differences between the Jovian planets and the terrestrial planets?

6.   Explain the most widely accepted theory for the formation of our solar system?

7.   Compare and contrast comets, asteroids and meteorites.

8.   List and describe the various types of meteorites?

9.   Characterize the internal structure and atmosphere of each of the planets of the solar system.

10. How did the layers of the Earth form?

11. Compare and contrast the interior structure of the moon to the Earth.

12. Contrast the geology of the mare and the highlands of the moon.

13. Explain the most widely accepted theory for the formation of the moon.

COMPLETION QUESTIONS

1.   The universe began between _____ and _____ billion years ago.

2.   The theory to explain the origin of the universe is called the _____.

3.   At the time of the Big Bang, matter did not exist; the universe was pure _____.

4.   Over time the percentage of hydrogen and helium in the universe has decreased from ____ to ____ percent.

5.   After the Big Bang the heavier elements were produced by nuclear _____ and distributed throughout the galaxy by the explosion of _____.

6.   The four closest planets to the sun are called the _____ planets and the next four are called the _____ planets.

7.   Our solar system formed from a rotating disk-shaped cloud of gas and dust called a _____.

8.   As the solar system evolved, most of the matter in the nebular cloud condensed in the center to form the _____, but local eddies in the disk accumulated cold matter to form _____.

9.   Between the terrestrial and Jovian planets is a belt of _____.

10. Ubiquitous _____ on the surface of many of the planets and of our moon indicate an early history of the solar system characterized by abundant collisions.

11. The _____ meteorites compose ____ percent of all meteorites and are composed primarily of silicate minerals.

12. The terrestrial planets are characterized by having a _____ core and a _____ mantle-crust.

13. Outgassing during volcanic eruptions on the _____ planets created their _____.

14. The smallest terrestrial planet is _____ which is also the one closest to the sun.

15. The thick atmosphere of Venus is composed of 96% _____.

16. The Jovian planets are characterized by having a _____ core.

17. The largest planet in the solar system is _____.

18. Much of the information we have on Uranus and Neptune we gained from the mission of _____.

19. _____ and _____ are characterized by huge storms which manifest themselves as large visible spots (red in one case, dark in the other).

20. _____ and _____ give off over twice the energy that they receive from the sun.

21. As the Earth and other terrestrial planets developed they heated up due to gravitational compression, impact of _____, and _____ decay.

22. As the early Earth heated and became more liquid, heavier elements sank to the center and lighter elements concentrated toward the outside in a process called _____.

23. As a result of this differentiation, the Earth has a silicate rich _____, and iron and magnesium silicate _____, and an iron-nickel rich _____.

24. While the Earth is _____ billion years old, the oldest crustal rocks we have yet found are _____ billion years.

25. The celestial body which we know the most about is _____.

26. The surface of the Moon can be divided into two parts: the low-lying dark-colored plains called the _____, and the higher, light-colored _____.

27. The highlands of the moon are primarily composed of _____ while the maria have a surface composed of _____.

28. The crust makes up about ____ percent of the moon.

29. The moon formed about _____ billion years ago.

30. Many scientists think that the moon originated from the impact of a giant _____ with the Earth.

MULTIPLE CHOICE

1. What was before the Big Bang?
   a. cosmic dust
   b. the sun and a few hundred stars
   c. a vast empty void
   d. You cannot speak of time before the Big Bang so the question is absurd!

2. The "Big Bang" theory explains
   a. the origin of the solar system.
   b. the origin of the universe.
   c. the origin of the asteroid belt.
   d. the origin of the moon.

3. The first two elements formed after the Big Bang were
   a. hydrogen and oxygen.
   b. hydrogen and helium.
   c. helium and oxygen.
   d. methane and helium.

4. It took _____ for the four basic forces, gravity, electromagnetic force, and strong and weak nuclear force to separate after the Big Bang
   a. less than a second
   b. about 100,000 years
   c. about 100 million years
   d. about a billion years

5. The terrestrial planets have
   a. a low density and are composed of gases.
   b. a high density and are composed of gases.
   c. a low density and are composed of rock and metallic elements.
   d. a high density and are composed of rock and metallic elements.

6. Which of the following is NOT a terrestrial planet?
   a. Neptune
   b. Mars
   c. Venus
   d. Mercury

7. The Jovian planets are composed mainly of
   a. hydrogen, helium, oxygen, and silicates.
   b. hydrogen, nitrogen, carbon dioxide, and carbon monoxide.
   c. hydrogen, helium, ammonia, and methane.
   d. water, oxygen, ammonia, and methane.

8. Meteorites composed primarily of iron and nickel alloys are called
   a. irons.
   b. stony-irons.
   c. stones.
   d. irony-stones.

9. The planet which rotates upright, but in a direction counter to all the others is
   a. Mercury.
   b. Neptune.
   c. Earth.
   d. Venus.

10. The smallest planet in our solar system is
    a. Mercury.
    b. Pluto.
    c. Mars.
    d. Krypton.

11. Earth is unique among the planets in having
    a. life.
    b. liquid water.
    c. an atmosphere dominated by nitrogen and oxygen.
    d. all of the above

12. Compared to the terrestrial planets the Jovian planets have
    a. greater density and greater mass.
    b. greater density but lower mass.
    c. lower density and lower mass.
    d. lower density but greater mass.

13. The largest planet in our solar system is
    a. Saturn.
    b. Neptune.
    c. Uranus.
    d. Jupiter.

14. The planet with the best known rings is
    a. Jupiter.
    b. Saturn.
    c. Mars.
    d. Venus.

15. The planet that rotates on its side is
    a. Mercury
    b. Venus
    c. Uranus
    d. Neptune

16. Compared to the solar system the Earth is
    a. younger.
    b. older.
    c. about the same age.

17. The heat that caused the early Earth to go into a molten phase came from
    a. gravitation contraction.
    b. meteoritic impact.
    c. radioactive decay.
    d. all of these

18. The densest part of the terrestrial planets is found in their
    a. cores.
    b. mantles.
    c. crusts.
    d. atmospheres.

19. The regolith on the surface of the moon originated from
    a. meteorite impact.
    b. chemical weathering of an atmosphere that is no longer present.
    c. cosmic dust in space.
    d. none of these

TRUE OR FALSE

___1.   Oxygen is the most abundant gas in Earth's atmosphere.

___2.   Nuclear fission formed the heavier elements from the lighter elements.

___3.   Earth is located near the center of the Milky Way Galaxy.

___4.   The asteroids probably formed from a planet between Mars and Jupiter which exploded.

___5.   Most of the material from the solar nebula, that formed our solar system, is found in the planets.

___6.   Unlike its neighbors, Earth has managed to escape large-scale meteorite impact.

___7.   The number of meteorite impacts on most planets has steadily increased over the past 4 billion years.

___8.   The terrestrial planets all have an early history of volcanism, cratering, and differentiation of their components.

___9.   Of all the planets, Venus is most similar to the Earth in size and density.

___10.   There is evidence of wind on Mars.

___11.   The largest known volcano in the solar system is Olympus Mons on Mercury.

___12.   Pluto is the only planet which does not orbit the sun on the plane of the ecliptic.

___13.   Earth is the only terrestrial planet with a moon.

___14.   The oldest continental rocks on Earth date at 4.6 billion years.

___15.   As a result of heating the heavier elements in the earth sank to form the core; and the lighter elements rose to form the crust and mantle.

___16.   The moon rotates on its axis one time for every trip around the Earth.

___17.   Saturn is the only planet in our solar system with rings.

___18.   The maria lavas are composed of basalt.

___19.   A popular theory is that the Moon may have formed by material ejected when a large planetesimal struck Earth early in its formation.

___20.   Just like the Earth has earthquakes, the Moon has moonquakes.

APPENDIX

ANSWERS TO QUESTIONS

Answers to Chapter Concept Questions: Because a primary goal of a study guide is to encourage study we are not providing detailed answers to these questions. To do so would encourage memorization of material that should be studied and understood and would, therefore, be counterproductive. A student should study the material in the text and formulate his/her own answers. We have, however, provided page numbers and figure and table references so that the student can readily find information necessary to answer the question. In some cases some additional hints are provided.

Answers to Tables: Filling in the tables in these questions is an excellent way to study some of the details of the chapters that many instructors expect students to know. Therefore, we are not providing detailed answers to these tables. Once again we are providing page numbers and table references where the pertinent information can be found.

Answers to Completion Questions: The answers provided are the preferred answers for the blanks in the questions. Answers for questions with multiple blanks are separated by a "/". Answers to these questions are given in the same order in which the blanks occur in the questions, but in some cases different order might be acceptable such as when the blanks are simply a list. The student should be able to tell when that situation occurs. In some cases more than one answer is possible and we have included those alternative answers in parentheses.

Answers to Multiple Choice and True or False Questions: The correct answers are provided.

Answers to Drawings and Figures: Because drawing talents vary so widely we are not providing detailed answers to these questions. To do so would take up too much space and might discourage those with limited drawing skills. In some cases you are simply asked to identify or label the features in a photograph or drawing. Depending on the complexity of the question either an answer or a page number or a figure reference is provided.

CHAPTER 1

Chapter Concept Questions:

1. Table 1-2
2. p. 9-11
3. p. 11-12
4. p. 13-14
5. p. 14
6. p. 16
7. p. 16-18
8. p. 15, Fig. 1.14
9. p. 14-16
10. Table 1.3
11. p. 19-20
12. p. 21-23

Tables: See page 15 for the pertinent data for both tables

Completion Questions:

1. minerals / rocks
2. layered rocks
3. deformation
4. fossils
5. hypothesis / theory
6. 4.6
7. solar nebula
8. sun / planetesimals
9. convection / mantle
10. away from

Multiple Choice: 1. d; 2. b; 3. d; 4. d; 5. d; 6. c; 7. d; 8. b; 9. d; 10. d; 11. d; 12. c; 13. c; 14. d; 15. c; 16. b; 17. c

True or False: 1. T; 2. F; 3. T; 4. F; 5. F; 6. T; 7. T; 8. T; 9. F; 10. F; 11. T; 12. F; 13. F; 14. F; 15. F

Drawings and Figures:

1. Fig. 1.17
2. Fig. 1.13
3. Fig. 1.15

CHAPTER 2

Chapter Concept Questions:

1. p. 31, 35-37, 39
2. p. 32
3. sugar and water are not, ice is, p. 31, 35-37
4. p. 32-33
5. carefully note the definition of mineral, p. 31, 35, Fig. 2-6
6. p. 33-34
7. p. 33-35
8. p. 35-36
9. p. 39
10. p. 37, 39
11. p. 44-47
12. p. 35,44
13. p. 46
14. p. 47
15. p. 47
16. p. 48-49

Completion Questions:

1. naturally / crystalline
2. Rocks
3. matter
4. elements / atoms
5. nucleus / protons / neutrons
6. atomic number
7. protons / neutrons
8. protons / neutrons
9. bonding
10. compound
11. ion
12. ionic / covalent
13. metallic
14. native elements
15. ions
16. oxygen / silicon / silicates
17. dark
18. carbonate

19. physical
20. color
21. cleavage / fracture

22. hardness
23. equal / water
24. rock / silicate

Multiple Choice: 1. d; 2. c; 3. b; 4. c; 5. c; 6. c; 7. c; 8. b; 9. d; 10. b; 11. c; 12. c; 13. a; 14. c; 15. b; 16. c; 17. d; 18. c; 19. c; 20. d

True or False: 1. F; 2. T; 3. T; 4. T; 5. T; 6. T; 7. F; 8. F; 9. F; 10. T; 11. T; 12. F; 13. F; 14. T; 15. F; 16. T; 17. T; 18. F; 19. T; 20. F; 21. T

Drawing Question
Fig. 2-11

CHAPTER 3

Chapter Concept Questions:
1. p. 57, 62
2. p. 58-59
3. p. 59
4. p. 59
5. p. 60-61
6. p. 62, 63
7. p. 63, Fig. 3.31

8. p. 66
9. p. 66-67
10. p. 68-69
11. p. 69-71
12. p. 73-74, Fig. 3-28
13. p. 74-75, Fig. 3-27, 3-29

Table: Table 3.1, pages 63-67, and Fig. 3.13 all contain useful information

Completion Questions:
1. extrusive / intrusive
2. calcium / sodium
3. olivine / biotite
4. 25
5. mafic / felsic
6. aphanitic
7. phaneritic
8. porphyritic
9. vesicular
10. pyroclastic or fragmental
11. texture / composition
12. 65
13. rhyolite
14. diorite
15. Basalt

16. olivine
17. pegmatite
18. tuff
19. granite
20. Obsidian / pumice
21. pluton (intrusion)
22. dike / sill
23. laccoliths
24. necks
25. 100
26. granitic (felsic) / dioritic (intermediate)
27. forceful injection
28. Stoping

Multiple Choice: 1. d; 2. d; 3. a; 4. d; 5. d; 6. b; 7. c; 8. c; 9. b; 10. c; 11. a; 12. b; 13. d; 14. a; 15. a; 16. d; 17. d; 18. b; 19. b; 20. a; 21. c; 22. c

True or False: 1. F; 2. T; 3. F; 4. F; 5. F; 6. T; 7. F; 8. F; 9. T; 10. F; 11. F; 12. T; 13. F; 14. T; 15. T; 16. T; 17. T; 18. F; 19. T; 20. F; 21. T; 22. T; 23. T

Drawing Question
1. This is a porphyritic texture. Its origin is explained on p. 62-63.
2. Perspective 3.2, p. 72

3. a. rhyolite is aphanitic, granite is phaneritic; b. basalt is aphanitic, gabbro is phaneritic; c. andesite is aphanitic, diorite is phaneritic.
4. Fig. 3.22

CHAPTER 4

Chapter Concept Questions:

1. p. 84
2. p. 84-86
3. p. 86-87
4. p. 86-88, 89, 92, 98, 99
5. p. 89-90
6. p. 93-95
7. p. 99-102, especially Fig. 4.22
8. p. 102
9. p. 103, Fig. 4-24

Completion Questions:

1. dormant / extinct (inactive)
2. water vapor
3. pressure ridges
4. columnar joints
5. pillow lava
6. ash / lapilli
7. volcanoes / fissures
8. crater
9. caldera
10. shield / fluid (mafic)
11. Kilauea
12. cinder
13. composite
14. lahars
15. shield
16. composite
17. basalt plateaus
18. pyroclastic sheet deposit
19. mid-ocean ridges
20. subduction zones
21. composite
22. hot spot

Multiple Choice: 1. a; 2. b; 3. b; 4. d; 5. c; 6. a; 7. c; 8. a; 9. c; 10. b; 11. c; 12. c; 13. a; 14. a;15. b; 16. b; 17. d; 18. c; 19. a; 20. d

True or False: 1. T; 2. F; 3. F; 4. F; 5. T; 6. T; 7. F; 8. F; 9. T; 10. F; 11. T; 12. T; 13. T; 14. T;15. F; 16. F; 17. F; 18. F; 19. F; 20. F; 21. T

Drawing Question

1. Fig. 4.12
2. Fig. 4.13, 4.16
3. Fig. 4.24

CHAPTER 5

Chapter Concept Questions:

1. p. 112-113
2. p. 113
3. p. 114
4. p. 114-115
5. p. 115
6. p. 116-117, Fig. 5-11
7. p. 116-117, 120
8. p. 121
9. p. 121, Table 5-1
10. p. 125-126, Fig. 5.19
11. p. 127
12. p. 129-131

Completion Questions:

1. weathering
2. erosion
3. differential
4. chemical / mechanical
5. frost
6. pressure / eroded

7. water
8. solution / oxidized / hydrolysis
9. feldspar / clay
10. Small (Clay)
11. faster
12. soils
13. organic
14. residual
15. A / B

16. accumulation / leaching
17. climate
18. pedalfer
19. caliche / pedocals
20. laterite / bauxite
21. deeper (thicker)
22. wind / water
23. residual
24. supergene

Multiple Choice: 1. c; 2. b; 3. c; 4. c; 5. d; 6. c; 7. b; 8. b; 9. c; 10. b; 11. c; 12. d; 13. c; 14. b; 15. b; 16. c; 17. d; 18. b; 19. d; 20. b; 21. a

True or False: 1. F; 2. T; 3. T; 4. F; 5. T; 6. T; 7. F; 8. F; 9. T; 10. T; 11. T; 12. F; 13. T; 14. F; 15. T; 16. F; 17. F; 18. T; 19. T; 20. T; 21. F; 22. F; 23. T

Drawing Questions
1. Figs. 5.2, 5.7, 5.5
2. Fig. 5.18
3. Fig. 5.25

CHAPTER 6

Chapter Concept Questions:
1. p. 140
2. p. 142
3. p. 145
4. p. 146
5. p. 146-147
6. p. 146-147, Fig. 6-11
7. p. 147
8. p. 147-148

9. p. 148-149
10. p. 149, Figs. 6.15, 6.16
11. p. 150-151
12. p. 153
13. p. 154
14. p. 156
15. p. 158-159, Figs. 6-28, 6-29

Tables: Tables 6.1 and 6.2 and pages 144-149

Completion Questions:
1. weathering (disintegration)
2. detrital
3. water
4. abrasion
5. sorting
6. deposited
7. lithification
8. compaction / cementation
9. clasts / clastic
10. 2 / gravel
11. rounding
12. quartz
13. claystone (mudrock)
14. fissility
15. coquina

16. Ooids
17. evaporates
18. chemical / quartz (silica)
19. Coal / oxygen
20. lignite / bituminous / anthracite
21. transgression / regression
22. structures
23. deposition
24. Cross-bedding
25. mold / cast
26. Coal / oil / natural gas (uranium)
27. source / reservoir
28. 50
29. Precambrian

Multiple Choice: 1. c; 2. b; 3. a; 4. a; 5. b; 6. d; 7. c; 8. d; 9. b; 10. a; 11. c; 12. d; 13. a; 14. a; 15. c; 16. a; 17. b; 18. d; 19. c; 20. c

True or False: 1. F; 2. F; 3. T; 4. T; 5. T; 6. T; 7. T; 8. F; 9. F; 10. F; 11. T; 12. F; 13. T; 14. T; 15. F; 16. T; 17. T; 18. F; 19. T; 20. T; 21. F; 22. T; 23. T

Drawings and Figures
1. Fig. 6.18, left to right
2. Fig. 6.5
3. Figs. 6.15, 6.16
4. Figs. 6.20, 6.21, 6.22

CHAPTER 7

Chapter Concept Questions:
1. p. 167
2. p. 168
3. p. 168-169
4. p. 170
5. p. 170-172
6. p. 171
7. p. 171-173
8. p. 173
9. p. 173-174
10. Table 7.1
11. p. 174-177, 180
12. p. 180-181
13. p. 182-183

Tables: Table 7.2 and pages 174-180

Completion Questions:
1. mineral / texture
2. shields
3. pressure / temperature / fluids
4. magma / geothermal
5. Lithostatic
6. differential
7. increases
8. pore spaces / magma / dehydration
9. contact / regional / dynamic
10. Contact
11. aureoles
12. Dynamic
13. regional
14. differential
15. Slate
16. Gneiss
17. Migmatites
18. calcite / limestone
19. quartzite
20. contact
21. anthracite
22. facies
23. isograds

Multiple Choice: 1. d; 2. b; 3. c; 4. d; 5. a; 6. b; 7. b; 8. b; 9. a; 10. c; 11. d; 12. c; 13. b; 14. c; 15. a; 16. b; 17. a; 18. a; 19. d; 20. b

True or False: 1. F; 2. F; 3. T; 4. T; 5. F; 6. F; 7. T; 8. T; 9. T; 10. F; 11. F; 12. F; 13. T; 14. F; 15. F; 16. T; 17. T; 18. T; 19. F; 20. T

Drawings and Figures
1. (a) sillimanite or kyanite and sillimanite both; (b) andalusite
2. Fig. 7.19
3. a. blueschist; b. sanidinite, amphibolite, greenschist, zeolite; c. granulite, amphibolite, greenschist, pumpellyite or zeolite

CHAPTER 8

Chapter Concept Questions:

1. p. 190-191
2. p. 193
3. p. 193
4. p. 194
5. p. 194-195
6. p. 195-197, 202-203
7. p. 197-199
8. p. 203
9. p. 203, Fig. 8.18
10. p. 208, 210
11. p. 208-212
12. p. 212
13. p. 212-213
14. p. 213, Fig. 8.26

Completion Questions:

1. Relative
2. Absolute
3. 4.6 billion
4. processes
5. bottom
6. original horizontality
7. younger
8. older
9. unconformities
10. angular
11. sedimentary / igneous / metamorphic
12. fossils
13. key beds
14. wide (large) / short (small, limited)
15. radioactive (parent) / daughter
16. protons / two
17. neutron / proton / electron
18. half life
19. mass spectrometer
20. sedimentary
21. 70,000
22. nitrogen-14 / beta
23. rings

Multiple Choice: 1. c; 2. d; 3. a; 4. c; 5. b; 6. a; 7. b; 8. b; 9. d; 10. a; 11. c; 12. d; 13. d; 14. a; 15. b; 16. a; 17. a; 18. d; 19. b; 20. c

True or False: 1. F; 2. T; 3. T; 4. F; 5. T; 6. T; 7. T; 8. F; 9. T; 10. T; 11. T; 12. F; 13. F; 14. T; 15. T; 16. F; 17. T; 18. T; 19. T; 20. T

Drawings and Figures

1. Fig. 8.3 and text on page 194
2. Figs. 8.9, 8.10, 8.11
3. Figs. 8.12, 8.13 and the discussion which accompanies these on pages 200-202

CHAPTER 9

Chapter Concept Questions:

1. p. 226
2. Fig. 9.5
3. p. 227-228, Fig. 9.6
4. p. 229-231, Figs. 9.7, 9.8
5. p. 232
6. p. 232, Fig. 9.13
7. p. 234, Fig. 9.15
8. p. 234, Fig. 9.16
9. p. 235, Table 9.2
10. p. 235, 238, Fig. 9.19
11. p. 239, 242
12. You will have to do some thinking here. Take notes as you read p. 239-244 and develop your own ideas.
13. p. 242-247
14. p. 246-248
15. p. 246, Fig. 9.29
16. p. 249-251

Completion Questions:

1. energy / elastic
2. elastic rebound
3. seismology
4. seismograph / seismogram

5. focus (hypocenter) / epicenter
6. 70 / 300
7. Benioff
8. circum-Pacific
9. Mediterranean - Asiatic / interiors / spreading
10. 900,000
11. body / surface
12. primary (P) / secondary (S)
13. primary (P) / parallel
14. Secondary (S) / perpendicular
15. Love (L) / Rayleigh (R)
16. primary (P) / secondary (S)
17. intensity / Mercalli
18. magnitude / Richter
19. 100
20. liquefaction
21. tsunamis
22. precursor

Multiple Choice: 1. c; 2. a; 3. a; 4. c; 5. a; 6. d; 7. a; 8. d; 9. b; 10. b; 11. c; 12. a;13. d; 14. d; 15 b; 16. b; 17. d; 18. d; 19. c; 20. d

True or False: 1. F; 2. T; 3. F; 4. T; 5. F; 6. F; 7. F; 8. F; 9. T; 10. T; 11. F; 12. F; 13. F; 14. F; 15. F; 16. T; 17. F; 18. F; 19. T; 20. T

Drawings and Figures
1. Fig. 9.8 and text on page 228
2. approx. 3500 km
3. about 3.6

CHAPTER 10

Chapter Concept Questions:
1. p. 260, 262
2. p. 262
3. p. 263-264
4. p. 264
5. p. 264
6. p. 265
7. p. 265-266
8. p. 266
9. p. 266-267
10. p. 267
11. p. 269-270
12. p. 271-272
13. p. 273-274
14. p. 274-275
15. p. 276

Tables: Table 10.2 and pages 264-266

Completion Questions:
1. composition / density
2. crust / oceanic crust / continental crust
3. granitic (felsic) / basaltic (mafic)
4. mantle / volume
5. core / iron
6. iron / nickel
7. seismic waves
8. density / elasticity
9. velocity / direction
10. density / elasticity /reflected
11. discontinuity / Moho (Mohorvicic discontinuity)
12. shadow
13. 103 /143 / focus
14. mineral structure
15. 25 °C / 1 °C
16. gravimeter
17. anomaly
18. isostacy / mantle
19. sink / rise
20. outer core
21. Curie point
22. polarity
23. inclination / declination
24. anomolies
25. paleomagnetism
26. normal / reverse

Multiple Choice: 1. a; 2. b; 3. c; 4. b; 5. d; 6. b; 7. c; 8. a; 9. d; 10. c; 11. b; 12. d; 13. d; 14. d, 15. d; 16. c; 17. c; 18. a; 19. c; 20. a

True or False: 1. T; 2. T; 3. F; 4. F; 5. F; 6. F; 7. F; 8. T; 9. T; 10. T; 11. T; 12. T; 13. T; 14. F, 15. F; 16. T; 17. F; 18. F; 19. T; 20. T; 21. T

Drawings and Figures
1. Fig. 10.2
2. Fig. 10.6
3. Fig. 10.9
4. Fig. 10.14

CHAPTER 11

Chapter Concept Questions:
1. p. 283-284
2. p. 285-287
3. p. 286-288
4. p. 288-289
5. p. 289
6. p. 290, Fig. 11.14
7. p. 290-292
8. p. 292
9. p. 292-293
10. p. 293, 296-297
11. p. 299-300
12. p. 300-301

Completion Questions:
1. 71
2. 3.5 billion / water vapor
3. *Glomar Challenger*
4. seismic
5. shelf / slope / rise
6. .1 / 4
7. canyons
8. turbidity / submarine fans / rise
9. abyssal
10. Active / trenches
11. shelves / rise
12. trailing
13. abyssal plains
14. trenches / 11,000
15. Pacific
16. 65,000
17. rift valleys
18. offset
19. seamounts / guyots
20. aseismic
21. pelagic
22. Ooze
23. Reefs
24. fringing / barrier
25. Ophiolites
26. Exclusive Economic

Multiple Choice: 1. d; 2. c; 3. d; 4. d; 5. a; 6. a; 7. c; 8. a; 9. d; 10. b; 11. a; 12. c; 13. b; 14. a; 15. d; 16. c; 17. c; 18. c; 19. c; 20. d

True or False: 1. F; 2. T; 3. F; 4. T; 5. F; 6. T; 7. F; 8. F; 9. T; 10. F; 11. T; 12. T; 13. F; 14. T;15. T; 16. T; 17. T; 18. F; 19. T; 20. F

Drawings and Figures
1. Fig. 11.22
2. Fig. 11.24

CHAPTER 12

Chapter Concept Questions:

1. 311-316
2. 316
3. 316
4. 316-318
5. 317-318, Fig. 12.12
6. 318
7. 320-321
8. 320-321, Fig. 12.16
9. 321
10. 321, 323-325, 327-328
11. 327-328
12. 328
13. 328, 331-332, Fig. 12.26
14. 332-333

Completion Questions:

1. Pangaea / Laurasia / Gondwana
2. *Glossopteris*
3. Alfred Wegner
4. South Africa
5. Sea-floor spreading
6. anomalies / mid-ocean ridge
7. 180 million / 3.96 billion
8. Deep-Sea Drilling
9. forms / destroyed (subducted)
10. lithosphere / asthenosphere
11. older / older / greater
12. basalt
13. rift valley
14. Benioff
15. intermediate
16. island arc
17. back-arc basin
18. subduction
19. melange
20. San Andreas
21. spots
22. Hawaii
23. convection

Multiple Choice: 1. b; 2. c; 3. b; 4. a; 5. b; 6. d; 7. c; 8. a; 9. d; 10. b; 11. b; 12. d; 13. d; 14. c; 15. b; 16. a; 17. c; 18. a; 19. b; 20. d; 21. d

True or False: 1. F; 2. F; 3. F; 4. T; 5. T; 6. F; 7. T; 8. F; 9. F; 10. T; 11. F; 12. F; 13. T; 14. T; 15. T; 16. T; 17. F; 18. F; 19. F; 20. F

Drawings and Figures

1. Fig. 12.15
2. Fig. 12.16
3. Figs. 12.19, 12.20

CHAPTER 13

Chapter Concept Questions:

1. p. 343
2. p. 343-344
3. p. 344, Fig. 13.15
4. p. 346-348, Figs. 13.9, 13.12
5. p. 348,350
6. p. 350-351
7. p. 362-354, Fig. 13.18
8. p. 354
9. p. 354
10. p. 355
11. p. 358
12. p. 358, 362-365
13. p. 365
14. p. 367
15. p. 367, Fig. 13.32
16. p. 367-370

Completion Questions:

1. force
2. compressional / tensional / shear
3. brittle / folds
4. strike / dip
5. anticline / syncline
6. axial plane / limb

7. overturned / recumbent
8. horizontal / inclined
9. away / toward
10. fault / joint
11. down
12. hanging wall / foot wall
13. reverse / 45
14. Shear
15. vertical / horizontal
16. system

17. normal
18. horsts / grabens
19. orogeny
20. accretionary wedge
21. ocean / continent
22. microplates
23. shields
24. Appalachian
25. Farallon

Multiple Choice: 1. b; 2. d; 3. b; 4. b; 5. a; 6. b; 7. a; 8. c; 9. d; 10. d; 11. c; 12. a; 13. b; 14. d; 15. a; 16. d; 17. c; 18. a; 19. d; 20. b; 21. a

True or False: 1. T; 2. T; 3. T; 4. F; 5. T; 6. T; 7. T; 8. F; 9. T; 10. F; 11. T; 12. T; 13. T; 14. F; 15. T; 16. F; 17. T; 18. T; 19. T; 20. F; 21. T

Drawings and Figures
1. Fig. 13.9
2. A. symmetric or upright anticline; B. symmetric or upright syncline; C. recumbent anticline; D. overturned anticline; E. overturned syncline
3. A. reverse fault; B. normal fault
4. right-lateral

CHAPTER 14

Chapter Concept Questions:
1. p. 379-381
2. p. 380-381
3. p. 381
4. p. 281-383, 385
5. p. 386
6. p. 386-391
7. p. 386-387

8. p. 387-388, Figs. 14.10, 14.12
9. p. 391-392
10. p. 395
11. p. 395-397, Fig. 14.21
12. p. 400-403
13. p. 400-403

Completion Questions:
1. shear / gravity
2. repose / 25 / 40
3. less
4. friction
5. parallel
6. earthquakes / water
7. rapid
8. rockfall
9. slump
10. rock slide

11. flow
12. mudflow
13. Debris
14. slowly
15. Quick
16. permafrost
17. creep
18. complex
19. slope stability
20. benching

Multiple Choice: 1. d; 2. d; 3. c; 4. d; 5. a; 6. a; 7. c; 8. c; 9. a; 10. c; 11. a; 12. d; 13. b; 14. c; 15. d; 16. b; 17. c; 18. c; 19. b; 20. c

True or False: 1. F; 2. T; 3. F; 4. T; 5. T; 6. F; 7. T; 8. F; 9. F; 10. T; 11. T; 12. F; 13. F; 14. T; 15. T; 16. T; 17. F; 18. F; 19. T; 20. T

Drawings and Figures
Figs. 14.12, 14.16, 14.21, 14.22

CHAPTER 15

Chapter Concept Questions:
1. p. 414-415, Fig. 15.5
2. p. 415. Fig. 15.6
3. p. 415
4. p. 416
5. p. 416-417
6. p. 417-418
7. p. 418, 420
8. p. 420-421
9. p. 421-422
10. p. 422-423, Figs. 15.18, 15.19, 15.20
11. p. 422, Figs. 15.18, 15.19
12. p. 424-425
13. p. 425, 428, Fig. 15.24
14. p. 429
15. p. 430-432
16. p. 432-434
17. p. 434, Fig. 15.30
18. p. 434-435
19. p. 435, Fig. 15.32
20. p. 436, Fig. 15.33
21. p. 436-438

Completion Questions:
1. 97.2 / 2.15
2. 20
3. Turbulent
4. infiltration capacity
5. sheet / channel
6. vertical drop
7. time
8. semicircle
9. friction
10. Discharge
11. 5 / kinetic
12. hydraulic
13. abrasion
14. bed / suspension / dissolved
15. Competency / capacity
16. channels / sediment
17. oxbow / meander
18. cut bank / point bar
19. floodplain
20. levee
21. standing (still) / decreases
22. progrades
23. distributaries
24. stream / bird
25. alluvial fan
26. drainage / divides
27. dendritic
28. radial
29. joints
30. base level
31. sea level
32. Graded
33. downcutting / headward erosion
34. base level
35. incised meanders
36. Superposed

Multiple Choice: 1. b; 2. c; 3. c; 4. d; 5. c; 6. a; 7. c; 8. c; 9. a; 10. d; 11. a; 12. a; 13. b; 14. d; 15. a; 16. d; 17. a; 18. b; 19. a; 20. a

True or False: 1. T; 2. T; 3. F; 4. F; 5. F; 6. F; 7. F; 8. T; 9. T; 10. F; 11. T; 12. F; 13. T; 14. T; 15. T; 16. T; 17. T; 18. F; 19. T; 20. F; 21. T

Drawings and Figures
1. 15.5
2. a. at least 30 cm/sec , 120 cm/sec, and 40 cm/ sec respectively. Note that the Y axis is logarithmic. These answers are, therefore, approximate.
3. Fig. 15.18
4. Fig. 15.26

CHAPTER 16

Chapter Concept Questions:
1. This is best considered and answered after reading the whole chapter but p. 445 makes some good points in the Introduction.
2. p. 445-446
3. p. 446
4. p. 446, Fig. 16.2
5. p. 446-447, Figs. 16.3, 16.4
6. p. 448, Fig. 16.6
7. p. 451, Fig. 16.9
8. p. 451-453, Fig. 16.12
9. p. 454
10. p. 454
11. p. 456-458, Fig. 16.17
12. p. 459
13. p. 459-460
14. p. 460-461
15. p. 462-463
16. p. 463, 465-468
17. p. 468

Completion Questions:
1. Porosity / permeability
2. aquifer / aquiclude
3. interconnected
4. water table
5. capillary fringe
6. gravity
7. high / low
8. water table
9. saturation
10. perched
11. saturation
12. depression
13. aquicludes / tilted / precipitation
14. artesian pressure
15. limestone / carbonic
16. sinkhole
17. stalactites / stalagmites / column
18. 30
19. saltwater incursion
20. pore
21. contamination
22. 37
23. magma
24. sinter
25. steam (gas)
26. geothermal

Multiple Choice: 1. d; 2. b; 3. a; 4. c; 5. a; 6. c; 7. b; 8. b; 9. b; 10. a; 11. a; 12. c; 13. b; 14. d; 15. a; 16. d; 17. b; 18. b; 19. c; 20. a

True or False: 1. F; 2. T; 3. F; 4. T; 5. T; 6. T; 7. F; 8. F; 9. F; 10. T; 11. F; 12. T; 13. F; 14. T; 15. F; 16. F; 17. T; 18. T; 19. F; 20. F

Drawings and Figures
1. Fig. 16.4
2. Fig. 16.17

CHAPTER 17

Chapter Concept Questions:
1. p. 477
2. p. 478
3. p. 478
4. p. 478-481
5. p. 481, Fig. 17.8
6. p. 482-483, Fig. 17.9
7. p. 482
8. p. 482-483
9. p. 485, Fig. 17.12
10. p. 488-489
11. p. 489-492
12. p. 492
13. p. 492-493
14. p. 493-496
15. p. 496-498
16. p. 491, 497, 498-499, 502, Fig. 17.28
17. p. 502
18. p. 502-504

19. p. 504
20. p. 504-505

21. p. 505

Completion Questions:

1. snow
2. one tenth / Australia
3. Antarctica / Greenland
4. 2.15
5. calving
6. sublimation
7. mineral
8. compacted / firn
9. 40
10. glacier
11. Plastic
12. friction / basal slip
13. continental / 50,000
14. Valley
15. 3000
16. accumulation / wastage / firn
17. snowfall / melting / sublimation / icebergs
18. retreat / advance / stagnant
19. accumulation
20. plastic / basal slip
21. brittle / crevasses
22. surge
23. centimeters / meters
24. erratics
25. Plucking / roche moutonée
26. polish
27. striations
28. U
29. fiords

30. hanging
31. cirque / arête
32. horn
33. continental / deranged
34. Canadian
35. drift
36. till / stratified
37. end / lateral
38. terminal
39. recessional
40. medial
41. drumlins
42. outwash / valley
43. kettle
44. eskers
45. kames
46. varves
47. year
48. dropstones / icebergs
49. 1.6
50. Four / 20
51. northern
52. pluvial
53. glacier / moraine
54. 18
55. 130
56. rise / 70
57. 300
58. rebound
59. Milankovitch

Multiple Choice: 1. a; 2. a; 3. b; 4. b; 5. c; 6. a; 7. c; 8. a; 9. d; 10. c; 11. b; 12. a; 13. b; 14. c; 15. c; 16. a; 17. b; 18. b; 19. d; 20. d; 21. c

True or False: 1. F; 2. T; 3. F; 4. T; 5. F; 6. T; 7. F; 8. T; 9. T; 10. T; 11. F; 12. F; 13. T; 14. F; 15. F; 16. F; 17. F; 18. T; 19. F; 20. T; 21. F; 22. F; 23. T; 24. T

Drawings and Figures

1. Fig. 17.15
2. Fig. 17.25
3. Fig. 17.28

CHAPTER 18

Chapter Concept Questions:

1. p. 514-515
2. p. 515
3. p. 515-516

4. p. 516
5. p. 517-518, Figs. 18.9, 18.10
6. p. 519

7. p. 519, 521-522, Figs. 18.12, 18.13, 18.14, 18.16
8. p. 522
9. p. 523
10. p. 524
11. p. 524-525

12. p. 525
13. p. 530
14. p. 530
15. p. 530
16. p. 531
17. p. 531-534

Completion Questions:
1. suspension / saltation
2. water
3. ventifacts
4. deflation
5. deflation hollow / desert pavement
6. shadow
7. windward / leeward
8. 30 / 34 / repose
9. barchan / parabolic
10. transverse / longitudinal
11. Loess
12. 10 / 30

13. 25
14. Coriolis / right / left
15. 20 / 30
16. rainshadow / east
17. Mechanical / temperature / frost
18. Rock varnish
19. internal / playa
20. fans / mouths / deposit
21. bajada
22. pediments
23. inselbergs
24. mesas / butte

Multiple Choice: 1. a; 2. b; 3. a; 4. d; 5. a; 6. b; 7. c; 8. a; 9. d; 10. d; 11. a; 12. b; 13. d; 14. d; 15. b; 16. b; 17. b; 18. d; 19. b; 20. c

True or False: 1. T; 2. F; 3. T; 4. T; 5. T; 6. F; 7. F; 8. T; 9. F; 10. F; 11. F; 12. F; 13. T; 14. T; 15. F; 16. F; 17. T; 18. T; 19. T; 20. F

Drawings and Figures
1. Figs. 18.12, 18.14, 18.13, 18.16
2. a. The vast white area is a salt pan in a playa and the valley is rimmed by alluvial fans
   b. pediment surface (Fig. 18.27)

CHAPTER 19

Chapter Concept Questions:
1. p. 542-543
2. p. 545
3. p. 545-546, Fig. 19.6
4. p. 545-547
5. p. 547
6. p. 547, Fig. 19.6
7. p. 547-548
8. p. 549
9. p. 549, Fig. 19.13
10. p. 549-550, Fig. 19.10

11. p. 550-551, Fig. 19.12
12. p. 552-553, Fig. 19.15
13. p. 553, 556, Figs. 19.16, 19.17
14. p. 557, Fig. 19.22
15. p. 558-559, Fig. 19.23
16. p. 559
17. p. 560, 564, Fig. 19.24
18. p. 562-563
19. p. 563-564

Completion Questions:
1. low / storm
2. flood / ebb
3. spring / neap
4. spring / neap
5. wind

6. crest / trough / height
7. rogue wave
8. circular
9. wave base / one half
10. seas

11.	swells	23.	quartz
12.	fetch	24.	spits
13.	breakers	25.	baymouth
14.	decrease / decrease / increase	26.	tombolo
15.	refracts	27.	parallel / lagoon
16.	parallel	28.	rivers / longshore
17.	Rip / sea	29.	erosion
18.	beach	30.	platform
19.	storms / berms	31.	seastack
20.	foreshore	32.	emergent
21.	drift	33.	submergent
22.	groins	34.	terraces

Multiple Choice: 1. b; 2. a; 3. d; 4. c; 5. d; 6. a; 7. d; 8. b; 9. b; 10. a; 11. c; 12. b; 13. d; 14. d; 15. b; 16. c; 17. a; 18. d; 19. b; 20. c; 21. d

True or False: 1. F; 2. F; 3. T; 4. T; 5. F; 6. F; 7. T; 8. T; 9. T; 10. F; 11. F; 12. T; 13. F; 14. T; 15. F; 16. T; 17. F; 18. F; 19. T; 20. F; 21. T

Drawings and Figures
1. Fig. 19.6
2. Fig. 19.12
3. Figs. 19.16, 19.17, 19.21, 19.24

CHAPTER 20

Chapter Concept Questions:

1.	p. 571-572, 587	8.	p. 575
2.	Table 20.1 is most concise	9.	p. 579-580, 582-587. The figures on these pages are very helpful.
3.	p. 571-572		
4.	p. 573-574	10.	p. 587, 589
5.	p. 574	11.	p. 589-591, Figs. 20.16, 20.18
6.	p. 574-575	12.	p. 590
7.	p. 575	13.	p. 590-591

Completion Questions:

1.	15 / 20	16.	rock (rocky)
2.	Big Bang	17.	Jupiter
3.	energy	18.	Voyager 2
4.	100 / 98	19.	Jupiter / Neptune
5.	fusion	20.	Jupiter / Saturn
6.	terrestrial / Jovian	21.	meteorites / radioactive
7.	solar nebula	22.	differentiation
8.	sun / planetesimals	23.	crust / mantle / core
9.	asteroids	24.	4.6 / 3.96
10.	craters	25.	the Moon
11.	stony / 93	26.	maria / highlands
12.	metallic / silicate	27.	anorthosite / basalt
13.	terrestrial / atmosphere	28.	12
14.	Mercury	29.	4.6
15.	carbon dioxide	30.	planetesimal

Multiple Choice: 1. d; 2. b; 3. b; 4. a; 5. d; 6. a; 7. c; 8. a; 9. d; 10. b; 11. d; 12. d; 13. d; 14. b; 15. c; 16. c; 17. d; 18. a; 19. a

True or False: 1. F; 2. F; 3. F; 4. F; 5. F; 6. F; 7. F; 8. T; 9. T; 10. T; 11. F; 12. T; 13. F; 14. F; 15. T; 16. T; 17. F; 18. T; 19. T; 20. T